WEATHER

WEATHER

How It Works and Why It Matters

ARTHUR UPGREN

JURGEN STOCK

PERSEUS PUBLISHING
Cambridge, Massachusetts

Library of Congress Catalog # 00-105193
ISBN 0-7382-0294-0

Perseus Publishing is a member of the Perseus Books Group.

Find us on the World Wide Web at http://www.perseuspublishing.com

Perseus Publishing books are available at special discounts for bulk purchases in the U.S. by corporations, institutions, and other organizations. For more information, please contact the Special Markets Department at HarperCollins Publishers, 10 East 53rd Street, New York, NY 10022, or call 1-212-207-7528.

Text design by Jeff Williams
Set in 11-point Minion by Perseus Publishing Services

First printing, August 2000
1 2 3 4 5 6 7 8 9 10—03 02 01 00

To Silvia, Jeanette, Bernhard, Frances, Josephine, and Eckhard
And to Joan and Amy

ACKNOWLEDGMENTS

The authors are grateful to Sally Brady and Amanda Cool for their help in making a book out of an idea. John Wareham has provided many of the figures, and Gabriele Zinn has found and secured some of them. We wish also to acknowledge the assistance of and discussions with Maria Avendaño, Paul Garmers, John Griese, Erhard Konerding, John Lee, Suzanne O'Connell, Linda Shettleworth, and Captain Rolf Wittmer, whose descriptions of El Niño were illuminating during a delightful cruise with him of the Galapagos Islands.

Contents

ILLUSTRATIONS

PREFACE

BY THE SEVENTEENTH CENTURY, experiment and observation were in full swing. Science had cast off the stultifying classical authority and burgeoned into the modern age.

Meteorology was one of the disciplines that benefited from this movement. The instruments by which we measure and quantify weather were developed; Galileo Galilei invented the thermometer and his student, Evangelista Torricelli, the barometer. With these and the hygrometer to measure humidity, Galileo, René Descartes, and others began to take readings and keep records.

In 1714, Gabriel Fahrenheit fashioned the temperature scale that still bears his name, and thirty years later, Anders Celsius did the same, but with a different scale and zero point. Researchers learned to equate certain swings in temperature or pressure with other atmospheric phenomena, and slowly the science of weather forecasting came about.

Gradually records of these quantities were scanned and averaged to form the concept of climate. Two trends occurred at more or less the same time over the latter half of the nineteenth and early twentieth centuries. One was the organization of weather stations into a network providing uniform coverage across much of Europe and the Americas, and later the rest of the world, replacing the collection of data accumulated by only a few dedicated individuals such as Benjamin Franklin. The other was a warming trend moving much of the world out of the relatively cold period known as the Little Ice Age that had begun in the fifteenth century into a milder period we like to think of as normal. The simultaneity of the two events might be thought to bias weather records in some fashion, but sufficient external evidence is available to prove that they happened independently of each other.

As meteorologists have come to realize, the one way in which twentieth century weather appears to be quite abnormal is in its very "normality". We may have been living in a time of comparatively modest extremes in temperature and precipitation. A substantial body of evidence points to worsening climate, to greater extremes in temperature and in rainfall. Droughts appear to alternate with floods more today than they did just a few decades ago, and summers are getting hotter as winters cool or remain the same. This is becoming the central concern of the study of climate, the weather data as averaged over the years and decades. The science of climatology is among the fastest growing disciplines in college curricula and in funds available for its advancement.

Climates vary widely in this world of ours. Indices have been established that divide them into groups of locales with similar conditions. Thus we have spatially differentiated climates, distinguished by latitude, availability of water, and other factors.

We are learning that climatic regimes change with time as well as location—satellites and computers are used to measure temporal change with ever-improving precision. As the human species moves toward making major alterations in climate, so has it become aware of these changes and the economic and other dislocations they are likely to generate.

These two trends may or may not be coincidental. But an awareness of them is becoming an essential part of a liberal education and a career in business or government. The authors intend to increase that awareness in an amicable way; the first seven chapters investigate the properties of weather, and the remaining chapters take a look at climates of the past, present, and as far as possible, the future as well.

INTRODUCTION

We have not inherited the World from our ancestors, we
have borrowed it from our children.

NATIVE AMERICAN SAYING

THE STORM INTENSIFIED THROUGHOUT THE NIGHT; from intermittent wind gusts it heaved itself into a raging turmoil. Off the east coast of the United States, no less than three low-pressure vortices, or *lows,* converged into a fury. The tide was high and the sea swept landward far beyond its normal range. The waves, whipped by hurricane-level winds, ripped at houses and high-rise buildings near the shore, tearing many of them to pieces in minutes. This storm had no equal.

Outside our suburban Philadelphia flat, the air pressure dropped. As the wind rushed past the outside wall, the consequence of a well-known natural law, the *Bernoulli Theorem,* set in. (This natural law has the same effect on an airplane wing; the resulting difference in pressure above and below the wing creates lift.) As the rising velocity of the air outside created a lower pressure, the pressure difference between the outside and inside wall surfaces of our building continued to build with the wind's velocity. The wind whined, its varied pitch a descant of parallel voices.

I heard a shattering crash. The bathroom window smashed out of the wall; glass, frame, casement and all went careening into the street. I was now part of the action as I scrambled to grab the papers blowing everywhere; some were even being sucked out through the jagged hole in the bathroom wall. Only much later in the night did this behemoth of a storm move out to sea and leave us in relative peace.

The storm of March 1962 was not a hurricane. Not at all. It did not form in the tropics and the season was wrong. But it behaved like a hurricane: It packed a hurricane's punch and caused extensive damage along the

Delaware and Maryland coasts. Winds and wave action reduced entire blocks of homes to rubble. The juxtaposition of three lows in conjunction with high tides raised one of the worst storms in the East Coast's history, not unlike the powerful and deadly storm Sebastian Junger described in *The Perfect Storm.*

This was a storm, then, full of sound and fury; but it faces an opposite number on the weather spectrum; this weather steams languidly upon the sky's stage, as it did for me one summer day in Vienna. It shows no rage, nor does it sing in a crashing, whining chorus. In recent years during the summer, a heavy and uniform yellow-gray sky often glowers darkly on an afternoon; only the Sun's diffuse rays shining through the haze breaks the oppressiveness. As we endure the pervasive, stifling heat and intense headache-breeding glare, no breath of air breaks the stillness; no zephyr relieves the miasma; no cloud breaks the uniform sky. When the Sun sets, no one sees it; the dull orange-red ball vanishes into the grayness while still well above the western horizon. Thus do we describe the most common of all types of weather, especially in or near urban regions. As the twentieth century drew to a close, this desolate weather pattern grew ever more common during the summers. It affected not only cities known for oppressive summer heat—Washington, Rome, and Athens—but also Vienna, Copenhagen, and London, cities that during the nineteenth century and most of the twentieth century were refuges from intense heat.

We have described two extremes in the weather of the temperate regions of the world. In the first, the winds of winter storms wreak havoc in barbaric proportions; in the second, not even a gentle susurrus stirs the leaves wilting in the unrelenting heat. Climatologists know that both these extremes are growing more common. Extreme though they appear, they constitute the "normal" weather of the future. Most of the twentieth century saw a gentle period of weather that was abnormal in its very normality and relative calm. The floods and drought, the long steamy summers and cold snowy winters are our likely lot for the coming century, and our adding greenhouse gases can only further and prolong the misery.

The swirling highs and lows, the warm and cold fronts, and the storms and calm interludes are normal features of the atmosphere of a rotating planet. Even if we could somehow smooth these irregularities into an undisturbed clear and even state, it would take only a few weeks for the air to return to its present diverse condition. Earth, with its endless procession of sunshine, overcast skies, and sunshine again, is unique among the objects in the solar system. Our marbled blue-and-white globe appears half clear and half cloudy from space. Ours is the only planet from which the Sun and the stars are visible on clear days and nights, but not on cloudy ones.

The other worlds in our solar system either hide behind impenetrable veiled atmospheres or bare their surfaces to the universe. In the first group we find Venus and the four gassy giant planets: Jupiter, Saturn, Uranus, and Neptune, along with Saturn's largest satellite, Titan, and, in a sense, the Sun, which is gaseous throughout. The terrestrial planets have desolate, barren surfaces covered with craters; these were formed in the early days of the system, when debris collided with debris everywhere before, during, and after the larger, gassy planets had coalesced and formed. Our globe is not interchangeable with any other; we must care for it if only because there is no other worthy real estate around.

No part of our natural environment is as visible and as variable as the weather. Neither the solid surface of our world nor the oceans undergo the variety and rapidity of change that occurs in the atmosphere. The atmosphere is the gaseous part of our planet and as such it is subject not only to a number of influences but to a rapidity of change that liquids and solids rarely experience. Together with the topography and composition of the surface, the atmosphere forms the background for a diversity of life, which it helps sustain. Through its normal diurnal and seasonal variation, the atmosphere defines the limits of agriculture, a discovery that, perhaps more than any other, led us from lives as nomadic hunter-gatherers to settled lives in villages and cities. A stable supply of food, in turn, allowed some citizens to take up trades not related directly to the accumulation of food: People became priests and scribes (and eventually, even meteorologists), developments that led to our modern complex society.

Climatic conditions once dominated human activity and to some extent still do; what natural sequence of events provokes more praise or vexation than the weather? Extreme weather conditions may cause disruptions in climate, but vegetation soon recovers, as does animal and human life. Primitive people learned to adapt to the varying conditions of their surroundings, but as civilization became more complex, with its globally intertwined economies, people were not always well prepared to survive during extreme weather conditions.

The state of our preparedness is the key to our future. Evidence for imminent major changes in the global climate is substantial and growing. Chief among these is a change, increasing at an ever faster rate, in the composition of the Earth's atmosphere. Most, if not all, of the changes result from human activity. The recognition of this fact has only recently brought climatology into the forefront of scientific study the world over. In a parallel development, the public in many countries is aware (through the media) of such terms as *greenhouse effect* and *ozone depletion,* which describe the effects of human activity on the atmosphere. Acid rain is another side-effect of civilization. As soon as acid rain was discovered and understood, it was attributed to human action. Although measures are now taken to arrest the influence of acid rain—the control of emissions from the exhausts of automobiles and industrial smoke stacks among them—it is thought to have caused the devastation of the forests in Europe and North America as well as the rapid and deplorable deterioration of such treasures as the great cathedral of Cologne and the Coliseum in Rome.

This is not the first time the Earth has undergone a significant climatic change. The ice ages are just one striking example of major natural atmospheric changes. More than ever before, the study of weather and climate, past and present, is important in gauging the climate of the future and the time scale over which changes are likely to occur. If the alterations in weather patterns result in detrimental consequences, it will be necessary to evaluate and correct those ascribed to human activity. We will have to make difficult choices, which the public may be reluctant to accept. One purpose of this book is the presentation of the facts.

Climatological research is now under way in many parts of the world, and every year more scientists focus on this subject. More research centers and laboratories appear along with newly designed and constructed equipment. Along with meteorology, the other earth and space sciences of astronomy, geology, and oceanography are being studied to determine the interrelations between them and our atmosphere. Despite the hurried flight of the astronomical establishment from planetary and stellar astronomy to the extragalactic realm, many scientists have noted the near correspondence of these four sciences to Aristotle's original four elements: earth, air, fire, and water.

In this book we will demonstrate how the natural sciences have become involved in climatology, and we will share with you some of the most spectacular results. Again, we emphasize that our purpose is to bring the interested reader up-to-date on the rapidly changing aspects of climate research and on the wide and broadening involvement of other sciences, thus enabling him or her to form enlightened opinions on the vexing problems before us.

We intend to document, among other things and as best we can, the latest cases for and against the dire predictions of global warming and the degree to which we should prepare for its continuation. Have we made the case beyond a reasonable doubt for worldwide concern and if so, to what extent should governments plan seriously for alternatives to our consumption of fossil fuels and their attendant liberation of carbon dioxide and other greenhouse gases into the atmosphere? We firmly believe that global warming ought to be an issue overriding normal party politics and petty bickering, both within the United States and between nation states and regions of the world.

•   •   •

It may be natural for many to regard such adverse conditions as the ozone hole and the greenhouse effect with skepticism. The oceans and atmosphere have always seemed to be of such immensity that they are beyond our ability to alter them for better or for worse. But the explosive growth of human population in the present century and the ability of recent tech-

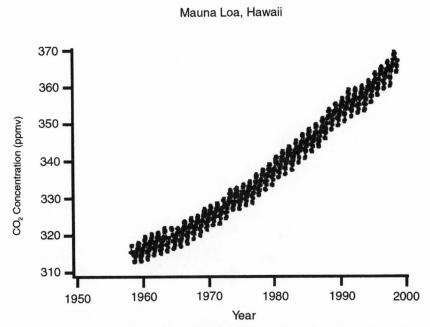

FIGURE 1.1   Increase in carbon dioxide at Mauna Loa from 1958 through 1998. (From C. D. Keeling and T. P. Whorf, Scripps Institution of Oceanography, University of California, La Jolla.)

nology to create devices that consume the resources of this planet at an ever faster rate have proved otherwise. The depletion of resources that took millions of years to form occurs so rapidly that we must reevaluate our present actions and their consequences for the world; such a reevaluation requires a complete understanding of the natural processes that affect the planet, especially its atmosphere.

Figure 1.1 shows the variation in the concentration of carbon dioxide in the atmosphere since 1958 as measured in Hawaii at the summit of Mauna Loa. Note the clear secular trend upward and the annual seasonal variations due to seasonal growth variations in the Northern Hemisphere. Since 1958, the amplitude of the annual variation has increased by 20 percent in Hawaii and by as much as 40 percent in more northerly latitudes. Figure 1.1 evokes the spirit and substance of this book; we intend to relate it to the possible consequences of its continued increase and suggest ways to avert such a crisis.

2

TEMPERATURE AND PRESSURE: THE FUNDAMENTALS OF AIR

If you haven't measured something, you really don't know very much about it.

KARL PEARSON (ATTRIB.)

WEATHER CAN BE DEFINED AS a visible and tangible manifestation of the physical conditions of the air at a given location and time, and of the changes in these conditions. Beyond that, we all know weather to be one of the most vexing, capricious, and fickle entities that we must endure. The atmospheric conditions at a given locality, specified by its geographic coordinates and elevation above sea level, can be numerically described by four measurable parameters: temperature, pressure, dew point, and precipitation. Of course, other expressions have also described weather, some intemperate, some vulgar, some charming, but we cannot transform these qualities into numbers. A general rule among weather people has it that those in the American west boast about their weather, but all others tend to knock theirs. In any event, we shall not proceed further with value judgments, no matter how justified they may be at times.

The Earth, like no other world, is a blend of matter in all three of its common forms: solid, liquid, and gaseous. These three states have very different properties. A rock is a solid object with a specific shape, and no meek effort will mold it into a different shape. But putty is also a solid, even if it doesn't appear to be much of one. The only difference between rock and putty is rigidity. If you squeeze a rock, it remains unchanged; but putty under gentle pressure quickly alters its shape. Imagine being struck on the head by rock or putty; the one would hurt at even a modest clout, but the other would require a fearful whack to do the same damage.

Liquid has an obvious difference; it flows to fill a swimming pool or any other cavity that contains it. Both liquid and gas fill the space available, but gas has no boundary at the top. Furthermore, unlike liquids, gases are compressible. Under only its own weight, the atmosphere compresses to the highest density at its bottom, at the solid or watery surface of the world. Water is just barely compressible and has only a slightly greater density at the bottom of the ocean than it has at the surface (a feature caused mostly by differing temperature). A slight degree of compressibility does exist, but for all but the most demanding of minutiae, water can be considered to remain at the same density no matter how much overlying material it is forced to bear.

We sense the near incompressibility of water in any number of ways. Try to stop the flow of water from a faucet or a hose. It just can't be done, at least not by hand, and reducing the size of the orifice through which the water gushes only acts to form a speedy jet-like squirt (as from a nozzle) and with a very high velocity. Only a powerful stopping mechanism, as is found in faucets, has the leverage to stop the flow.

Water reveals its refusal to compress in other ways as well. Diving into a swimming pool can be risky if one does not enter either head first or feet first. A "belly flop" presents a large and undignified cross section of the body to the water, which won't be pushed out of the way, and the skin around the belly stings as a result. As a consequence of the incompressibility, an object displaces the same amount and weight of water at any depth. Thus when the Titanic took on enough water to weigh more than the water it displaced, it went straight down to the ocean floor, sinking 12,500 feet in less than five minutes.

Air has no rigidity at all; it fills the space available for it, unless that space is large. Air and any of its constituent gases can be visualized as many tiny particles all moving about, bouncing into each other and into any walls or other limits comprised of solid or liquid matter, all subject to gravitational pull. The air temperature relates directly to the average velocity of the particles, a fact which helps to define it. When the number of particles is extremely large, as it always is in our ordinary experience, the distribution of the particles' velocities is predictable. Large numbers are customarily more workable than small ones: An insurance company's actuarial tables work well when millions of people form the available sample, but usually fail badly when only a few dozen are considered.

•   •   •

We can measure air, like almost anything else, and express its properties in numbers. Think of the types of air that might be surrounding you. Air can be hot or cold, moving or still, humid or dry, pure or polluted. Watch any televised weather forecast and you find measures for each of these varying properties. Consider heat. Just what is temperature and what causes a change from hot to cold?

Air turns out to be nothing more than zillions of tiny particles, called molecules, all rushing around and colliding into each other again and again. There are so many zillions of them in even a small room that the chances of their being distributed unevenly throughout the room rather than uniformly is vanishingly small. We suspect that readers can quickly imagine that if a hundred coins were tossed into the air, the chance that every one of them would come up heads would be incredibly small, about one in 1,000,000,000,000,000,000,000,000,000,000 (one followed by thirty zeros). There are so many more molecules of air in our room that the chance of their all moving one way, leaving a vacuum in another part of the room, is something akin to the chance that a chimpanzee picking away on a typewriter will type out all of *Othello* and *Henry IV* (Parts I and II).

So the temperature of a gas is the distribution in the speeds of these molecules as they rush around. Air is a mixture of molecules of several gases. These molecules are not all the same weight. The two most common components of our air are nitrogen and oxygen, which between them account for 99 percent of our atmosphere. They weigh 14 and 16 atomic units respectively, where 1 atomic unit is about the weight of a hydrogen atom. The two lightest gases, present only in trace amounts, are hydrogen and helium. They are lightweights, weighing in at 2 and 4 units, whereas big, sluggish carbon dioxide tips the scales at 44 units. Molecules are like football players. Who wants to bang into a 300-pound lineman? Better a 200-pound running back, or best of all, a 120-pound cheerleader if colliding is called for. Small molecules bounce off any of the big boys much as most of us would bounce off a sumo wrestler. We would do the bouncing, whereas the quarter-ton sumo wrestler would move little, if at all. In that same sense, the light molecules dance around faster than the heavier ones.

Heat, then, is a measure of the distributions of the velocities of air particles. Next time you feel the intense summer heat, remember that air molecules are hitting you faster than usual. And when winter cold chills you to the bone, move to where those air particles move faster. With cold, the speeds lessen, and at very cold temperatures they become downright sluggish. There is a limit to temperature at which none of the particles move at all; we call this limit *absolute zero,* beyond which temperatures cannot exist and have no meaning. Absolute zero is never quite reached in nature or in the laboratory, although it can be closely approached in both. This limit to cold occurs at about −273 degrees on the Centigrade or Celsius Scale, conveniently abbreviated *C* in either case, or −459 degrees on the Fahrenheit Scale, identified with the letter *F.*

Above this point, the speeds of gas molecules can rise steadily to define the temperature. For this reason, it is convenient to introduce another scale with zero at this coldest point, but with the same size degrees as in the Celsius Scale. Here, the melting and boiling points of water occur at +273 and +373 degrees respectively, instead of 0 and +100 degrees as in the Celsius, or +32 and +212 degrees as in the Fahrenheit. Simply add 273 to the Celsius temperature to determine it on the new system, called the Kelvin Scale and represented by the letter *K.* The concept "twice as hot" has a real meaning on the Kelvin Scale: A temperature of 400 degrees is twice as hot as one of 200 degrees, and the gaseous particles move twice as fast. Other temperature scales have been devised, but none survives in common use. Figure 2.1 shows the three scales.

Air has weight. This gives it the properties of momentum and inertia, but they aren't apparent to our senses unless the air is moving, in which case we feel wind. How much does air weigh? Water weighs just about the same amount everywhere. On the British system, water checks in at 62.4 pounds per cubic foot. In the metric system, the density of water translates the unit of volume, the cubic centimeter, into the unit of weight, the gram. By definition, a gram weighs 1 cubic centimeter of water. By contrast, 1 cubic meter (1 million cubic centimeters) of air at sea-level pressure weighs 1,293 kilograms, or almost 3,000 pounds; 1 cubic foot of air weighs 1.3 ounces. Water thus weighs 773 times as much as an equivalent volume of air at sea level (and at a temperature of 0°C, or 32°F).

A lot of air packs a lot of weight, illustrated when one compares the weight of air to that of the Eiffel Tower. That web-like structure is 300 meters (984 feet) high; its four legs form a square 100 meters (328 feet) on a side. The Eiffel Tower weighs about 6,500 tons, whereas the air in a cylinder just large enough to surround and enclose it weighs more; it scales just over 6,700 tons.

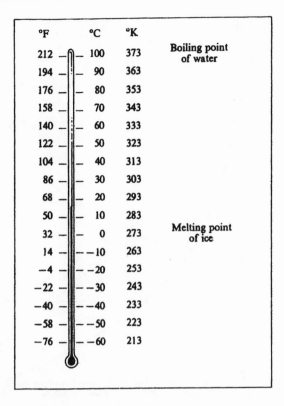

°F	°C	°K	
212	100	373	**Boiling point of water**
194	90	363	
176	80	353	
158	70	343	
140	60	333	
122	50	323	
104	40	313	
86	30	303	
68	20	293	
50	10	283	
32	0	273	**Melting point of ice**
14	−10	263	
−4	−20	253	
−22	−30	243	
−40	−40	233	
−58	−50	223	
−76	−60	213	

FIGURE 2.1   A comparison of three temperature scales.

Unlike water, air is compressible—very much so. A simple bicycle pump illustrates that. And just as a pump compresses the air under force, so does the atmosphere compress under the force of its own weight. But that air is not infinitely compressible can be useful, even lifesaving. Elevator shafts, now built to just fit the elevator, do not permit a free fall should the car break loose from its supporting cables. In July 1945, when a bomber struck the Empire State Building in New York City at its seventy-ninth floor, an elevator with its cables sliced fell at such a gradual rate that its operator survived a drop of over 900 feet.

Imagine an atmosphere like ours, but made of incompressible air. At sea level, the air would remain as it is, but it would not get thinner with increasing altitude. In that event, our atmosphere would form a layer just about 5 miles (8 kilometers) thick, the distance being defined in this manner as the *scale height*. Above a height of 5 miles would be empty space, nothing more. But air packs down under the weight of more air above it,

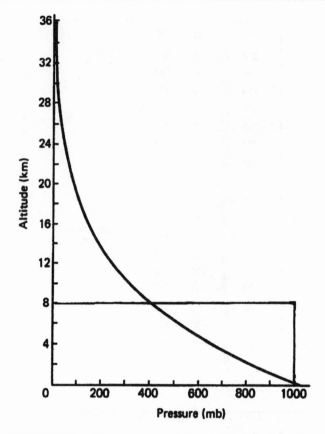

FIGURE 2.2   Air pressure variation with altitude and
scale height.

so in reality, the density falls off in a specific way. We don't even know exactly where the top of the atmosphere is. At some point, hundreds of miles
up, the molecules of the atmosphere become so thinly spaced that we can
think of it as a vacuum for all practical purposes. Figure 2.2 shows the actual case and the scale height of our atmosphere.

The weight of a column of air one inch square and extending upward
into space, or to the 5-mile scale height in our imaginary incompressible
atmosphere, weighs about 14.7 pounds. That constitutes the air pressure at
sea level. Because water weighs 773 times as much as air, a 1-square-inch
column of water need be only 1/773 of 5 miles, or 34 feet, to weigh the
same. For each 34 feet down into the ocean, the pressure adds one more
atmosphere's worth of weight. Thus at the 12,500-foot depth of the Ti-

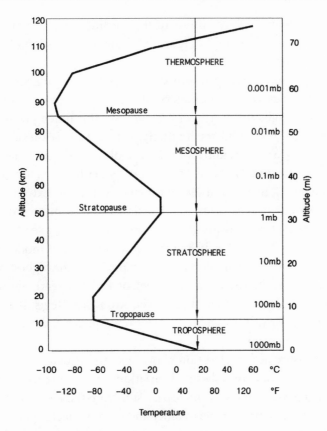

FIGURE 2.3   The variation of temperature with altitude.

tanic, the pressure is about 370 times that at sea level, around 2.72 tons per square inch. The bathyscaphes carrying motion picture director James Cameron and others down to view and film the wreckage had to be tough indeed to resist this pressure.

Air temperatures vary over a wide range, too wide a range to suit most of us. The lowest temperatures ever measured at the surface of the Earth are –68°C (or –90°F) at an inhabited site in Siberia, and –88°C (or –127°F) in Antarctica. The highest widely accepted surface air temperatures (in the shade) occurred in the Sahara desert in Libya, where a temperature of 58°C (or 136°F) has been recorded, and at the bottom of Death Valley in California, reaching values of 57°C (or 134°F) measured under standard procedures. The air temperature aloft is usually much colder than at the surface and depends strongly on the elevation. At 10,000 meters (33,000 feet), the altitude near which commercial jet airliners fly, the temperature is usually around −45°C (-49°F). Figure 2.3 shows the variation of temperature with altitude.

The average atmospheric pressure at sea level is 1,013 millibars, equivalent to 29.92 inches, or 76 centimeters, of mercury. The weight of the air directly above weighs this much and thus makes this much pressure. Values higher than 1,050 millibars are rarely seen, but the sea-level pressure may drop as low as 900 millibars in the eye of a hurricane. These limits correspond to around 31 inches, or almost 79 centimeters, on the high side, and below 27 inches and 69 centimeters at the low extreme. At the summit of Mount Everest (with an elevation of 8,850 meters, or 29,028 feet), the atmospheric pressure is about one-third that at sea level; two-thirds of the atmosphere lies below this altitude. Even at moderate altitudes, such as Mount Washington in New Hampshire (the highest point in the northeastern United States at 6,288 feet, or 1,917 meters), the air pressure is but 79 percent of that at sea level. Atop Ben Nevis, the highest mountain in the British Isles at 4,406 feet (1,343 meters), the air pressure is reduced to 85 percent.

The air pressure at any elevation above sea level is determined almost entirely by elevation, with only small variations in pressure over time due to the normal passage of highs and lows. Aside from these fluctuations, the pressure falls off in a predictable manner and no other factor plays such an important a role in any understanding of the vertical structure of the atmosphere. The variation of air density with elevation is close to that of the pressure with elevation, but it is not identical to it. Although this sounds redundant, it's not. The compressibility of air means that it is squished in proportion to the air above it, which is doing the squishing: more weight on top, more squeezing underneath. As we have seen, water doesn't do this—the Titanic lost no time on its way to the bottom. But a hot-air balloon seeks its level depending on its weight and that of the air it displaces.

IT'S NOT THE HEAT, IT'S THE HEAT AND THE HUMIDITY

The great tragedy of science—the slaying of a beautiful hypothesis by an ugly fact.

THOMAS HENRY HUXLEY

AIR IS A MIXTURE OF GASES, along with minute solid and liquid particles in suspension such as water droplets, customarily seen as clouds or fog. Air is not a compound, as is carbon dioxide (for example, carbon dioxide's molecules are identical and contain two atoms of oxygen in combination with one atom of carbon), nor is it a simple element such as oxygen, hydrogen, or even iron.

Our atmosphere consists primarily of two gases, nitrogen and oxygen, both of which exist in what is called *diatomic form* (two atoms per molecule, designated as N_2 and O_2). The majority (78 percent) is nitrogen; oxygen makes up another 21 percent. The remaining gases amount to only 1 percent of the total, and most of that is argon, one of the inert gases (gases that do not combine with other elements). Carbon dioxide accounts for just over 0.03 percent, and many other gases are present in trace amounts.

Such is the case for dry air, air without any water vapor. Water vapor is always present, but in widely variable amounts. In cold dry air, such as lies above Antarctica, the vapor content is as low as 0.1 percent of the air. Above a tropical rain forest, vapor may amount to as much as 3 percent. If we could separate these gases and stack them up in constant-density amounts, nitrogen would account for about four miles of the scale height of five miles, and the layer of oxygen would be about one mile thick. Argon would form a layer of 75 meters (250 feet); the layer of carbon dioxide would be only about 8 feet (2.5 meters) thick. The remaining gases would amount to a layer of about 6 inches (15 centimeters).

The water vapor content of the air undergoes large variations from one time and place to another, and in the stacking picture varies from about 25 feet to almost 700 feet, or from 8 to 200 meters. Ozone, on the other hand, of which most is actually concentrated in the stratosphere, makes up only a total of about 6 centimeters. Yet we rely on that trace amount to ward off the harmful ultraviolet radiation from the Sun.

Carbon dioxide is essential for the sustenance of all forms of life. It is important to realize just how little of our air consists of this gas, because carbon dioxide is much the most important and abundant of the gases that give rise to global warming, the so-called greenhouse gases that produce the *greenhouse effect*. The greenhouse effect is the well-known condition that keeps a greenhouse warm in the wintertime and makes a car unseemly hot when in the sun with its windows closed. The glass in both cases is transparent to the incoming solar radiation, but opaque to the inevitable reradiation from inside the greenhouse or car. The solar energy occurs mostly as visible radiation with a maximum in the yellow region, but the car reradiates in the infrared, invisible to our eyes. Window glass is opaque to this radiation to the extent that the car heats up until some of the radiation can escape. Like a hot car, the atmosphere blocks the reradiation and heats up the surface.

Water can be present in the air in two forms: as water vapor, an invisible gas, or as minute droplets that float in the air and are seen as haze, fog, or clouds. Water vapor, the gas whose abundance varies greatly, plays a major role in any discussion of the atmosphere. One of the few absolutes in the science of meteorology is that warm air can hold more water in the form of vapor than can cold air—lots more. The maximum amount of moisture occurring in the air is fixed by its temperature; if too much vapor is present, *saturation* occurs and the excess vapor condenses out in the form of minute water droplets or ice crystals. A rise of 10°C (or 18°F) in the air temperature doubles the maximum amount of water that the air can sustain as vapor, although this ratio varies somewhat over the range of possible temperatures.

If we cool a blob of air, keeping the pressure constant, we will eventually reach a temperature at which the air is fully saturated and, if cooled further, can no longer hold its moisture. This lowest temperature is called the *dew point,* or the *dew point temperature,* and it must necessarily be as cold as or colder than the air temperature.

The Earth's atmosphere is divided into layers; the layer closest to the surface is called the *troposphere* and is a region of mostly falling temperature with altitude. Physically, the troposphere is characterized through its major energy input, which comes from the heated surface below; and the surface as an energy source makes impossible any widespread stable configuration. The troposphere is where all "weather" takes place, with clouds, precipitation, winds, and ever-changing global and local patterns. Because clouds, rain, and snow require relatively dense air for their formation, they are confined to low, dense layers of the atmosphere in any event.

The upper boundary of the troposphere is clearly marked as a temperature minimum and is called the *tropopause.* The tropopause varies in altitude from about 9 kilometers (6 miles) near the poles to nearly 16 kilometers (10 miles) above the Tropics. Just above the tropopause lies the stratosphere, a clear region in which the temperature rises with altitude, which produces stable stratification. The reason for this temperature rise is as follows: We noted above that cooling air loses its ability to hang on to its water vapor, and if cooled enough, the dew point is reached and condensation or precipitation occurs. Warm and cool air have one more fundamental difference: Warm air is always lighter and more buoyant than cool air and will rise whenever cool air surrounds it. The difference can be great enough to lift a balloon; indeed, balloonists depended upon this phenomenon throughout the nineteenth century. Thus, as sunlight warms the ground, a blob of air near it expands and rises, now called a *thermal.* Up the thermal goes, expanding and cooling along the way, until its density matches the temperature of surrounding air; if by that point the thermal has reached dew point, it becomes a cloud. If our thermal, our blob from near the surface, continues to rise, the cloud it has formed at dew point will expand vertically, occasionally to the top of the troposphere. There it stops. In the stratosphere, the temperature rises with altitude, a situation called a *temperature inversion.* Should our blob attempt to rise any higher, it would be surrounded by ever-warmer air, so this is a no-no. Inversions sometimes occur near the surface, too; the air goes nowhere but just sits there and, if the inversion occurs over a city, collects smog.

●   ●   ●

Several factors determine our degree of discomfort in the heat and the cold. Moisture, called *humidity,* is an important contributor to our discomfort. Most over-stress humidity's importance, as indicated by the phrase "It's not the heat, it's the humidity," now considered dogma by most of us in the United States. But what *is* humidity? Why do weather forecasters and news people disdain it and frequently delete it from their telecasts unless ordered not to by management?

To understand the complex role of water vapor in the atmosphere, we must understand humidity as an expression of discomfort. Humidity is not an absolute quantity; it is relative, defined as the ratio of vapor present to that required to produce saturation for a given temperature.

Relative humidity (popularly known as just humidity) and dew point temperature are the two most common means to describe the wetness of air. Neither is a precisely measurable quantity at very low levels, but the dew point carries one of the important tests for any parameter: Unlike humidity, dew point is an independent quantity. The humidity is determined from the existing air temperature and the dew point temperature, always as cool as or cooler than the air temperature. The problem with the relative humidity is that, because it is not a fully independent parameter, it depends on the value of the temperature. As an illustration, we consider the height and weight of a person to be two parameters of that person, but these parameters are not two independent quantities; one is partly a factor of the other. Tall people are on average heavier than short people; a person weighing 250 pounds is usually considered fat, but many basketball players weigh this much or more and no one would think of them as fat. A seven-foot-tall man weighs 250 pounds, even if he is quite thin. Weight is dependent on (and partly predicted from) height.

In the same manner, the temperature and the humidity are not independent measures of discomfort; one is partially a function of the other. This is why meteorologists would like to abandon humidity and use only the dew point, which is independent of the temperature and therefore gives more information than does the humidity. Mark Twain is said to have commented that there are three kinds of lies: lies, damn lies, and statistics. The manipulable humidity is the sort of statistic that gives statistics a bad name!

Thirty years ago, a single index of personal discomfort was created, defined by the temperature and the dew point, but not the humidity. Several forms arose, their differences slight; one was defined as equal to 0.4 times the sum of the two temperatures plus 15 degrees; this was called the *DI*, the *discomfort index*. After the predictable howls of protest issued forth from chambers of commerce and others, it was rechristened the *THI*, the

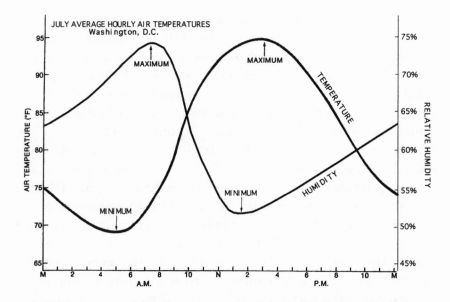

FIGURE 3.1   The variation of temperature and humidity on a typical hot summer day.

temperature-humidity index. Even this was too much for many places that push their climates as dry, even if insufferably hot, and the whole thing was allowed to pass into oblivion. Now we are once again allowed to believe that relative humidity is the only important part of our discomfort.

But wait! This relative humidity often varies by as much as 50 percent on a sunny day at any one location without any infusion of air with different properties. It can be a sticky 90 percent at one time of day and down to a refreshing 30 percent just a few hours later. If we compare various locations at various times of day, we could find a rain forest climate drier than a desert climate. On a typical summer's day, the temperature and humidity perform a kind of dance back and forth, as we show in Figure 3.1. In midafternoon at the hottest time of day, the humidity has dropped to a daily low, and just before dawn, the usual coldest moment, the humidity can be high to the point of saturation. All the while the air remains the same air. If a dry front passes through, the two curves may be a little farther apart, that is all.

Although we cannot measure accurately the low levels of humidity or dew point that occur in dry air, here is a useful way to compare the heat discomfort of one place to another: On the weather page of a daily newspaper glance at the lowest temperature reported in the last twenty-four hours and recall the formula for the DI or THI. For a humid city such as Washington, D.C., the daily high and low temperatures in July might be 95° and 75°F (35° and 24°C), with a dew point at 70°F (21°C), not much below the low for the day. At Phoenix, the two might be 110° and 70°F (43°and 21°C) or, with this city's rapid population growth, abundant irrigation, and many swimming pools helping out, 80°F (27°C) for a low. Washington's DI is thus 81 and Phoenix's is 87 (or 91, if the latter low is used). The DI was arranged so that at 70, no one feels uncomfortable, and for each point above 70, 10 percent feel uncomfortable. Thus a DI of 74 means that four people out of ten feel hot, and at 80, everyone feels hot. From the examples above, everyone in both places is miserable, but some people are more miserable than others.

Many factors prevent the DI from revealing whether someone is comfortable or miserable or somewhere in between. Activity, location in the sun or shade, clothing, a breeze, a cool drink—all these affect the situation.

In the wintertime, the situation reverses. The wind speed brings in the factor called *wind chill.* Special tables show a wind chill temperature dependent upon the air temperature and the wind speed and the resulting colder temperature one feels. Wind chill has become a mainstay of televised weather reports, probably because it provides more sensational weather news than does the temperature alone. But, like humidity, wind chill is often misleading, and it is an even more capricious variable. Anyone walking on the streets of midtown New York (or any other big city) quickly realizes that the skyscrapers funnel the wind along the avenues or the cross streets (and occasionally both) because they act as canyons between the tall buildings. Turn a corner, and the wind chill can change by as much as thirty or forty degrees. Just where, then, should the wind-chill temperature be measured and reported?

Most major weather stations are located at airports, but the problem of measuring wind-chill temperature exists there, too. Winter wind is commonly gusty and the wind chill flops all over as a result; this condition is reported in good faith by the well meaning, but its use is obviously limited.

Although a wind in summer decreases one's discomfort, it is surprising that a wind-chill effect is never reported in the warmer months. In both extremes of temperature, summer and winter, the effects of humidity and wind chill get mixed together, and this confuses many into an incorrect estimation of both of them. People confusing humidity with wind chill speak of a dry cold in winter, meaning that at one location the air temperature really doesn't feel all that bad when compared to another. Having resided in Cleveland, Minneapolis, New York, and Washington, we can say with certainty that, on average, a temperature of zero feels the same in all four cities. There are calm days in all four cities when the cold seems less uncomfortable against the bare skin, but the average wind speed as well as the average humidity is the same at the same air temperature. Discomfort is a personal matter, and susceptible to personal prejudice, particularly when one bases one's experience on a limited time in a city: "I spent a weekend in Washington, D.C.—and it was the hottest and most humid place I've ever experienced." This pronouncement may bring a modicum of satisfaction to the speaker, but it is not based on objectivity or science. Factors such as humidity, wind chill, and summer heat indices help to jazz up weather telecasts, but they don't describe the state of the weather as satisfactorily as such independent measures as temperature, dew point, and pressure.

· · ·

On the Earth, unlike any other planet, clouds are made out of water. This refreshing compound, known chemically as H_2O, exists in abundance in both liquid and gaseous forms on the surface of the Earth—but of no other world in the solar system. Were we to discover otherwise, the chances for life there would rise dramatically. One of Jupiter's four large moons, Europa, might reveal liquid water below its icy surface to a near-future space mission. But oceans of the stuff lying about for our use and enjoyment are not found elsewhere. Precipitation comes in several flavors: rain, freezing rain, sleet, snow, hail, dew, and frost. The last two, dew and frost, aren't strictly considered precipitation at all; they form on grass or the side of a glass of cold water or beer through condensation when the air in contact cools past the point of saturation and can't hold its moisture. We get dew (water) if the air is above freezing and frost (ice) if below. Depending on conditions, any single storm can produce rain, freezing rain, sleet, or snow, or a melange of any two or more. It is a matter of the temperature profile with altitude.

Precipitation as such comes out of clouds (or fog if the cloud is on or near the ground), regions of the atmosphere where the rising air has dropped its temperature to the dew point. Massive condensation sets in, either as droplets or as miniature ice crystals, depending on the temperature. The presence of such condensation nuclei as dust particles, salt crystals, and molecules of the oxides of nitrogen or sulfur helps the condensation a great deal. This stuff, whether manmade or natural, is collectively called *aerosols;* if it is manmade, it is known as air pollution as well. If no condensation nuclei are handy, the air may remain *supersaturated,* a term we will define later. Once droplets or ice crystals are present, their surfaces speed condensation and hence the growth of the droplets or crystals. Mergers of droplets or crystals help along further growth; this process is particularly effective in thunderstorm clouds, where differences of electrical charge between drops can make them merge almost violently.

The first droplets or crystals are small and light and float in the air. As they grow they begin to fall. They may start out as ice and melt on their way down when they reach warmer layers of air. If the temperature is cold enough all the way to the surface, the precipitation reaches the ground as snow, otherwise it will arrive as rain. But if very cold air close to the ground lies beneath warmer air, rain coming down from the upper warm layers may refreeze, reaching the ground in the form of small ice pellets known as *sleet.*

Finally, we come to freezing rain, perhaps the worst and most dangerous form of all. After a spell of very cold and dry conditions with temperatures well below freezing, trees and power lines and almost everything else outside are also at a temperature well below freezing. If warm and moist air then moves into the area, the water vapor freezes onto the trees and power lines, lining them with thick layers of ice; these layers may become so thick that the branches snap. Such an event (erroneously termed an *ice storm*) affected parts of Canada and Maine in January 1998; it disrupted power lines in places for more than a week and left millions in the cold.

Freezing rain of incredible proportions that all but leveled central Connecticut on December 17, 1973 was referred to by NBC News as the worst ice storm on record. Tree limbs, six inches thicker than they should be, cracked and fell; the resulting bedlam sounded like a guerrilla war. Live cables sparked in streets and around cars and most folks fled their dark, frigid homes.

Another sequence of events may also lead to freezing rain: Drops falling into a very cold layer very near the ground may refreeze at the last moment and reach the ground as sleet pellets or as icy rain.

Weather stations measure the water equivalent of the precipitation with a rain gauge into which the rain or melted sleet or snow falls. Strict procedures allow meteorologists to compare present and past records to detect climate changes. Liquid rain appears easy to measure, one needs only a rain gauge that is mounted well away from any overhanging tree branch or wall.

Measuring snowfall is much more difficult and can be accomplished only by following strict procedures. The standard and acceptable method of snowfall measure requires that a level surface be measured and then promptly cleared of snow every hour during the storm. Failure to observe this procedure, day and night, always results in a spuriously low register because snow compacts over time and appears reduced as a result.

Having said this, we must point out that these procedures cannot always be followed; wind frequently piles up the falling snow into uneven levels, called *drifts*. Where should one measure the snowfall, then? Whenever drifts are present, an average depth must be estimated, not just the highest or lowest point. All snow must be measured and reported to obtain a proper monthly or seasonal total; other forms of frozen precipitation (sleet and freezing rain) should also be included.

These procedures, and any involving climate assessment, are vital because any failure to observe them results in an inaccurate diminished measure. If some inaccurate procedures giving a higher total were offset by others lowering the total, the danger would be of less concern because some balancing out would result. But alas, anyone measuring snowfall on his own will find lower snowfalls than official records, and, perhaps of greater significance, gain a false impression that snowfall today doesn't match the averages of past years.

Measuring snowfall fuels one of the most mistaken beliefs in all of meteorology: that snowfalls are less than they used to be. During lectures, everyone raises his or her hand in response to the question, Were snowfalls higher when you were young? Even when you were a child and drifts came up higher because you were smaller, you darn well know they were still greater in the absolute! People of all times, now and earlier, believe this. A little reasoning shows that it could only be true if snowfall perpetually and monotonically diminishes every year or so. Records belie this in any location, but nothing can shake the belief, and sloppy recordkeeping only acts to reaffirm it. In New England, with good records lasting for centuries, snowfall amounts since colonial times are about as great in the last few decades as they were in any equivalent period in the past.

Unlike other species, people are notoriously adept at creating self-fulfilling prophesies. Psychologists call the effect of our remembering some observations but not others *one-sided recall.* One-sided recall leads, for example, to the widespread but false belief that crimes occur more frequently near the time of the full Moon. Careful observation shows no increase in crime at that time despite firm beliefs to the contrary. This is a case of *mumpsimus,* the continued belief and acceptance of a tenet proven wrong, of exposed but customary error. Mumpsimus is as common today as it was in medieval times, even if the subjects involved have mutated away from witches and spectral apparitions, and it seems to arise nowhere more often than in our recollections of past weather. Selective beliefs are pervasive and commonplace in our acceptance of what may or may not be true about the weather, and exposing them is the proper work of every educator.

The mistaken belief that many have about the abundance of snow may be reinforced because two properties of snowfall are customarily measured. One property is the amount of snowfall, whereas the other is the number of days snow covers the ground. People may confuse one with the other, and the two properties differ greatly. For one example, Minneapolis receives in an average winter only a little more snow than does Hartford. But on about twice as many days in Minneapolis the ground is covered with one inch or more of snow. The Midwestern city being much colder, its snow in takes much longer to melt; hence, the difference.

Irving Berlin's *White Christmas,* a song beloved even by those who have fled to the sunbelt, reveals a similar belief, namely, that Christmas seasons were white "then" but they are not "now." Again, the facts just don't support the belief. Over more than a century, the percentages of snow-covered Christmas seasons have remained the same. Perhaps idealized snow occurs in the rural scenes portrayed in Currier and Ives prints, whereas today's white stuff turns quickly into slush in the city streets.

•　•　•

When people began to measure and record meteorological parameters such as temperature, cloud cover, precipitation, and wind speed and direction in a systematic way, meteorology entered its modern phase. Record-keeping, even if only in the memories of a few keepers of the calendar, must have been around as long as agriculture so that people could correctly gauge the right times to plant and to reap.

Farmers may have been the first to keep a written record of their weather observations for their own use in future planting and harvesting seasons. All but one of these parameters could easily be expressed in some numerical form without the help of equipment; the exception is temperature, which could be expressed only in a qualitative way on the basis of personal impressions of heat or cold.

The first real thermometers, of which one has survived, were built in Italy in the seventeenth century, allegedly by Galileo. It was long known that liquids and solids expand when heated up and that the rate of expansion was different from one material to another. The first thermometer consisted of a glass container to which was welded a long thin glass pipe curled into a spiral with several turns. The pipe was linearly graduated, and the entire system filled with a liquid, probably water or alcohol. Many modern thermometers work on the same principle. Although these instruments were sensitive to the ambient temperature, they nevertheless lacked both a definition of the temperature scale and a means for calibration.

By this, we mean that a given and arbitrarily chosen value for the temperature must be assigned to a temperature condition that can be replicated, such as the temperature at which ice melts. Many temperature scales have been proposed and put into practice, all based on assigning values to two reproducible temperatures, the particular device interpolating between them. The scales proposed by Celsius and Fahrenheit are still used today.

Most scale designers chose for their two fixed points the temperatures of melting ice and boiling water. Difficulties arise, though, because the temperature of a mixture of water and ice depends critically on its salt content. Thus distilled water must be used if one wishes to control with accuracy the respective temperatures.

The establishment of the boiling point of water fares no better because it depends critically on the air pressure, which is obviously variable. To this we must add that the expansion coefficient for various types of glass and liquids is not strictly a linear function of the temperature; in addition, expansion coefficients differ from one material to another. Finally, as if that weren't enough, the width of the glass pipe through which the liquid expands must remain constant over the entire length of the thermometer.

All this means that it is very difficult to construct two thermometers that over their entire range will give identical temperatures. To obtain uniform data worldwide requires a previous comparison of any thermometer with an arbitrarily chosen standard thermometer.

The definition of intermediate temperatures between the two fixed points has to be based on the unsatisfactory combination of the expansion coefficients of two materials, glass and a liquid. For meteorological purposes, when only an intercomparison of temperatures at various sites and at various times is wanted, such a definition may be adequate; but it is inadequate when physics must be applied to the structure of the atmosphere. The theory of gases finally led to a satisfactory definition of the temperature scale based on the average motion of the particles that compose the gas. The scale for this physical temperature was chosen so that at the two previously used fixed points the physical temperature agreed with the temperature on the Celsius scale. Among other things, the theory led to the discovery that there is an absolute zero point at 273.2 degrees below the freezing point of pure water, as we pointed out in Chapter 2.

Another difficulty with the liquid-in-glass thermometers—filled normally with mercury but occasionally with alcohol—is that they are slow to register the temperature. The variation of the temperature has to penetrate the glass before it can heat or cool the liquid. Glass is a poor conductor of heat, or of anything else. Nowadays, electronic thermometers have a very fast response. Some of them can detect temperature variations occurring within one hundredth of a second. Furthermore, their high sensitivity enables them to detect variations of only one-hundredth of a degree.

Standardized liquid-in-glass thermometers of satisfactory accuracy became available in the latter part of the eighteenth century and were soon available in Europe and North America. Only a few official weather stations were around then, but many private individuals dedicated considerable time and effort to recording their local weather conditions. No one more vigorously sought a network of weather records at that time than Ben Franklin; he was among the first to detect the prevailing westerlies, the winds that direct most weather systems to the east in midlatitudes. Franklin's records can still be found in libraries and private collections.

Air temperature near the ground—we're talking here about the first ten to twenty meters—depends critically on two parameters: the height above the surface and the nature of the ground cover. During the day, particularly under sunny conditions, the temperature drops rapidly with the distance from the surface. At 10 meters above the ground (depending on the nature of the surface), the temperature may be several degrees lower than that only a few centimeters above the ground. In the early morning after a clear night, the opposite may be true; at that hour, the temperature at 10 meters above the ground is several degrees higher than it is closer to the ground.

It is not enough, then, to insert a thermometer into the air and read the temperature. For a reliable reading, one must also supply information on the height above the ground and the nature of the ground cover below the spot where the temperature was measured. In former times, thermometers were simply held at eye level, normally over a grassy area. Rarely was the ground condition recorded—whether the grass was wet and green or dry and brown; such facts can make a noticeable difference. Readings were sometimes taken from a roof, at a height different from the usual one, and over a different surface. Naturally, it was difficult to interpret such temperature measurements, made before strict standardization rules were formed and put into practice. Yet, unlike the measure of snowfall, where any mistake reduces the snow measured from the correct value, here to some extent we can average over the readings to establish a reasonably valid mean temperature.

A few weather stations have kept continuous temperature records for almost two centuries. Yet without information about the calibration of the thermometers and the local parameters, we can draw only limited conclusions from the data.

Other, more distant, surroundings can also affect the temperature; conditions in a large metropolitan area can be quite different from those of the open country. Concrete and asphalt absorbs the daytime *insolation* (a term shortened from *incoming solar radiation*) and reradiates it at night; thus the city becomes a heat island rather warmer than the surrounding countryside.

Most of the older weather stations were built in the middle of a large park in the center of a then-small town. Now they find themselves in the middle of a large city; long-term trends found in the temperature data of these cities may be the result of urbanization and have nothing to do with any other temperature trend. This heat island effect can raise a city center's temperature at night above that of the surrounding countryside by as much as 5 to 10 degrees.

The measurement of precipitation, at least in liquid form, appears less complicated. Almost any kind of receptacle can serve this purpose; however, some complications exist. Wind creates turbulence, and can alter the course of a raindrop or a snowflake just before it falls on the ground or into the rain gauge. Turbulence is created downwind from any obstacle, and the drops tend to fall right there. Even the rain gauge can be such an obstacle, deviating the drops away from it, as can nearby trees or buildings. The need for a clear area around the gauge has always been recognized, but only recently have aerodynamically constructed rain gauges been adopted. When we interpret long-term precipitation records, we must consider all these factors.

The accurate measurement of the velocity of wind requires sophisticated equipment. Furthermore, the wind's normally gusty and turbulent nature creates difficulties in arriving at some consistent and meaningful value of the velocity (recall that velocity implies speed and direction, not

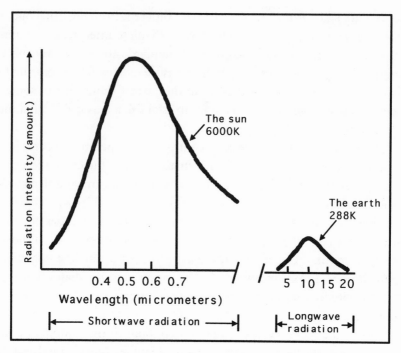

FIGURE 3.2   The distribution of radiant energy from the Sun and the Earth.

just speed alone). Wind speed and direction are greatly affected by local obstacles as well as by the topography of the surrounding area, and frequently vary considerably with the elevation above the ground. For these reasons, the Beaufort Scale was introduced. The Beaufort Scale is based on two sets of visible effects produced by the wind from which the wind speed can be estimated. One set applies to effects occurring on land, from slight motions of smoke, flags, or branches to the uprooting of trees and damage to buildings; the other observes the water surface from the presence of waves and whitecaps. But the reconstruction of even a crude history of the winds at any site is nearly impossible.

The cloud cover of the Earth has a decisive influence on the incoming solar radiation and on the outgoing surface radiation and consequently has a strong effect on the temperature. The reconstruction of the global cloud cover's history (if it were possible) would be useful. Clouds are an extremely variable feature of the atmosphere, both with time and with place. Before the satellite era, obtaining reliable information on the extent of the planet covered by clouds at any one time was difficult, and even more so on the types of clouds present. Such information is important because we have reason to believe that man's activity does indeed interfere with natural cloud formation. Unfortunately, reliable data of global coverage are available only for the years since 1970, too short a time span to detect a significant trend.

4

THE FOUR SEASONS

Sometimes too hot the eye of Heaven shines.

WILLIAM SHAKESPEARE, *SONNET XVIII*

ENERGY COMES FROM THE SUN—we all know that. But if we think about it, we find energy seemingly from other sources too, but that energy also comes from the Sun, even if indirectly. Water power, wind power, energy from fossil fuels (coal, oil, gas) all derived from our day star at one time or another.

Most of the physical processes that go on within the atmosphere are also initiated or caused by the Sun. Some part of the insolation is reflected directly back into the sky and the rest is absorbed by the air and by the surface. The part reflected, the reflective capacity of the surface, is known as the *albedo*. The albedo of a mirror would be about 100 percent, but an asphalt tennis court reflects only perhaps 5 percent or less of the incoming light and the rest is all absorbed and reradiated as heat, as anyone knows who plays on asphalt in the midday sun. Table 4.1 shows the albedos of various substances.

Clouds and snow reflect most of the sunlight, which is no surprise to a skier: On the slopes, glare from all that reflected sunlight forces the skier to use protective lenses, goggles, or sunglasses. In most of the countryside, the glare is much less than it is on the ski slopes because trees and fields reflect less incoming light and heat; at a beach, sunlight reflected from the water or the sand creates more glare than a field or a street does, though not as much as snow.

Fortunately for us, the radiation emanating from the Sun is almost constant and has been so for millions of years. But the amount falling on the surface of the earth varies every minute because it depends on the angle of incidence of incoming sunlight. Radiation reaches a maximum when the

TABLE 4.1   Albedos of various substances

Substance	Average Albedo
Snow	75
Clouds	70
Ice	35
Sand	30
Open Water	15
Grass	15
Forests	15
Soil	10

Sun is overhead; it becomes insignificant at or near sunrise or sunset (as is shown in Figure 3.1). In addition to the diurnal variation of the Sun's altitude from dawn to noon to dusk, a pronounced seasonal variation is present in the insolation, at least for high and intermediate geographical latitudes. The variation is modest at the equator and greatest at high latitudes as a result of the extreme seasonal variation in those regions. Even so, variation caused by clouds can be considerable: As clouds become grayer, less insolation reaches the ground.

•   •   •

The year is customarily divided into four seasons. But how are they marked? Which days should belong to each season on the calendar? There are two ways of delineating the seasons, each serving a specific purpose. One, the common one, is astronomical in origin; the other is more meteorological.

The two methods use different approaches because the extremes in the mean monthly temperature do not coincide with the maximum amount of insolation. In the Northern Hemisphere, the maximum amount of daylight occurs around June 21, when the Sun is farthest north in the sky; but the hottest weather of the year usually occurs in July and August. Similarly, the shortest day is reached on December 21, but the coldest months are January and February. This effect is the well-known "lag of the seasons" and demonstrates the heat inertia of the combined system of atmosphere and oceans, due mostly to the enormous heat capacity of the oceans. Land heats up more quickly and cools off more quickly than does water; this is why Kansas City has hotter summers and colder winters than San Francisco has with the Pacific Ocean immediately to its west; both cities are at about the same geographical latitude. Sunshine on land penetrates into the ground only a few feet, if that. In water, as any diver knows, the light filters down several hundred feet; only below that level is it dark all the time, day and night. Water heats and cools more slowly than land because solar energy is spread over such a depth. Water is a good heat conductor, much more so than sand or rock, and thus conducts heat easily to greater depths.

If we look closely at the origin of holidays in the temperate midlatitudes, we can just detect the vestiges of an alternative division of the year into four equal parts. This archaic tradition harks back to prehistoric times, when many ancient peoples shifted from a nomadic hunter-gatherer existence to a settled agrarian one; this change began about 8,000 to 10,000 years ago.

In the temperate latitudes of the Earth, where most early civilizations arose, early farmers discovered that the lag of the seasons, which cause the hottest and coldest periods of the year, occurred well after the longest and shortest days by almost half a season, or a month and a half. This seasonal lag arrived later in maritime climates and sooner in the continental regimes. The hottest and coldest periods of the year are never very far from the first days of August and February, respectively. Halfway between these two dates, on or about the first of May and the first of November, fall the midway points that mark the beginning and the end of the warm half of the year. This is a climatic, or meteorological, way of dividing the year into four nearly equal seasons. Offset as they are by about half a season from their astronomical counterparts, they give rise to four additional important festival dates; these are roughly equally spaced throughout the year, subdividing it into eight nearly equal parts of about a month and a half each.

Thus the seasons are marked by what the Sun does, or by what the temperature does, or both. In cultures such as ancient Celtic society, all eight begin with holidays and festivals, some of which we still celebrate. Christmas, Candlemas or Groundhog Day, Easter, Mayday or Walpurgis, Midsummer Vigil or a nearby Independence Day, Lammas, Labor Day or Michaelmas or any other harvest days, and Halloween with All Saints Day all derive from dates along the eight-season division.

The seasons result from one major and two minor causes; the dominating cause is the one most of us were taught in school. Most globes are built to show a cant or tilt, that is, the axis of the globe leans over like the leaning tower of Pisa. Globemakers do know how to make a model that does

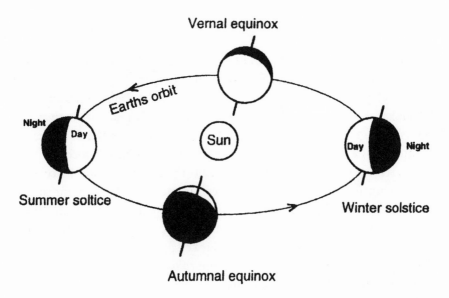

FIGURE 4.1   The seasons and their principal cause, the inclination of the Earth's axis.

not lean over, but they choose to include the lean, thus making science teachers think before they speak.

The reason for the leaning globes is the inclination of the Earth's axis with respect to the direction perpendicular to its orbital plane, which astronomers call the *plane of the ecliptic.* This angle amounts to about 23.5 degrees (varying slightly over the millennia). In June, the inclination tilts the Northern Hemisphere toward the Sun and the Southern Hemisphere away from it (see Figure 4.1). In December, the positions reverse. Taken together with the direction of the axis with respect to the direction towards the Sun and the latitude of the observer, the tilt determines the maximum angle of incidence of the solar radiation (that is, the altitude of the Sun above the horizon at noon) as well as the duration of daylight at a given latitude. In Australia, or "down under," the seasons are reversed. Christmas can be hot, with sunlight lasting far into the evening, and a displaced American may celebrate Independence Day in the snow.

This is all a natural consequence of the Earth's formation, condensing as it did into a ball that revolves around the Sun in the plane of the ecliptic—but rotates in a plane inclined by the 23.5-degree angle. We can state it another way: This plane is inclined by 23.5 degrees from the plane of the

Earth's equator extended onto the celestial sphere, forming the Celestial Equator.

The ecliptic defining the path of the Sun in the sky crosses the Celestial Equator at two points. The Sun, then, must cross the Celestial Equator twice each year. It does so on or close to March 21 and again on September 23, and at these times the periods of daylight and darkness are each twelve hours as seen from any point on the Earth. These points and dates are the equinoxes, the vernal equinox and the autumnal equinox, respectively. After March 21, the Sun appears to move northward away from the celestial equator until June 21, when it reaches the most northerly point on the ecliptic, called the summer solstice. The Sun then moves south for six months, passing through the autumnal equinox, until it reaches the winter solstice on about December 22. The four key dates, the two equinoxes and the two solstices together, are called the *colures,* and the passages of the Sun define the astronomical seasons.

In designating these points, we are Northern-Hemisphere chauvinistic; spring, after all, comes in September to lands south of the equator. But because most of the land and 90 percent of all people live north of the equator and inhabit the Northern Hemisphere, we will define them as above.

Other astronomical factors have a small but significant influence on the seasonal effect; they help us understand the climate history of our planet, and we will describe them later. One, related to the distance between the Earth and the Sun, is addressed here. Since 1609, when Johannes Kepler published his discovery that planetary orbits are elliptical and not circular, the Sun's distance has been known to vary from the average of about 93 million miles, or 150 million kilometers. The separation reaches a minimum in early January, about January 4 or 5, and a maximum six months later, in early July. The closest and farthest points along the planet's orbit are called *perihelion* and *aphelion,* and differ by about 1.5 million miles (2.5 million kilometers) on either side of the average distance. In these extreme positions, the Earth receives about 3 percent more and less solar energy in January and July, respectively, than it does on average. At present, the arrival of the Earth at perihelion happens in the wintertime in the Northern Hemisphere and summertime in the Southern Hemisphere, and the farthest point, the aphelion, is reached when the seasons reverse. This coincidence has the effect of intensifying the seasonal extremes in the Southern Hemisphere while moderating them in the Northern Hemisphere. Because most of the Southern Hemisphere is covered by water (unlike the Northern Hemisphere, which is dominated by land, as is ap-

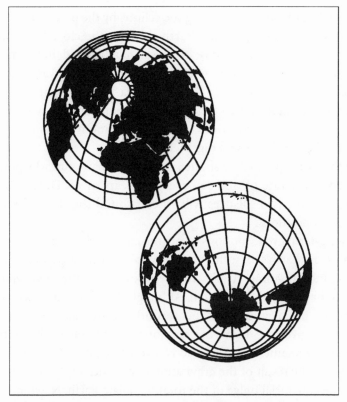

FIGURE 4.2 The Northern and Southern Hemispheres, showing the distribution of land and water.

parent in Figure 4.2), the intensification there is not pronounced because the thermal inertia of the oceans tempers the climate; that is, the oceans heat and cool more slowly than the land. As a result, maritime climates have less extreme temperatures than climates inland. The variation of distance from the Sun and the sizes and distribution of the continents and the oceans are the two minor causes of the seasons.

Kepler also discovered that planets move at their slowest near aphelion and fastest when nearest the Sun at perihelion, as required by the laws of motion and the force of gravity. The imbalance in time is quite noticeable in our calendar; the period from March 21, the beginning of spring, to September 23, the usual beginning of autumn, lasts 186 days, one week longer than the remaining part of the year, 179 days.

The times of perihelion and aphelion passages have not always taken place in January and July as they do now. As the equinoxes and solstices are carried backward (westward) along the ecliptic by the precessional motion (see below), the passage of the Earth through the closest and farthest points from the Sun will advance in the calendar. Indeed, in the middle of the thirteenth century, not long after the Magna Carta was created and Genghis Khan was terrorizing much of Eurasia, these two points coincided with the two solstices. At that time, the closest point and the winter solstice (when the Sun is farthest south) fell together about December 22, and their opposite numbers came together on June 22.

Now they are separated by about 14 days; each pair of nearby points has separated at the rate of one day every 50 years or so. During the next 21,000 years, the two solstices will wheel around the sky and return to their present positions.

Two long-term cyclic motions create the change; the first is a gyroscopic motion of the Earth's axis; this causes a wobble with a period just under 26,000 years. This slow, steady wobble is known as the *precession of the equinoxes*, or simply the *precession*. Although the Alexandrian astronomer Hipparchus discovered it in the second century B.C. (a remarkable achievement, given the degree of precision it takes to detect the motions of the stars), it remained for Sir Isaac Newton to explain the reason for it. Precession is the result of the gravitational influence of the Moon and the Sun on the equatorial bulge of the rotating oblate Earth. Newton saw it as identical to the motion of a spinning top, or gyroscope, as its axis starts to lean away from a vertical orientation. The Earth's gravity wants to pull the top over, but the Earth responds by precessing to the side. For the same reason, the Earth's axis also moves sideways. Precession causes many aspects of the sky, including the orientations of the constellations, to change slowly over the centuries.

As a result of the precession, the Earth will be tilted—its axis will be pointing—in the opposite direction 13,000 years from now from its location today near Polaris, the pole star at the present time (see Figure 4.3). This precession in turn will enhance the extremes of the seasons in the Northern Hemisphere and moderate them south of the equator, just the opposite of the case today. The great land masses of the Northern Hemisphere, Eurasia, North America, and most of Africa do not share the gentler maritime climates of their smaller southern counterparts (excluding uninhabited Antarctica), and the seasonal extremes in the north are expected to be greater than at present.

Although the true period of the precession is near 26,000 years, its effect on climate acts over a shorter one, closer to 21,000 years, as we mentioned

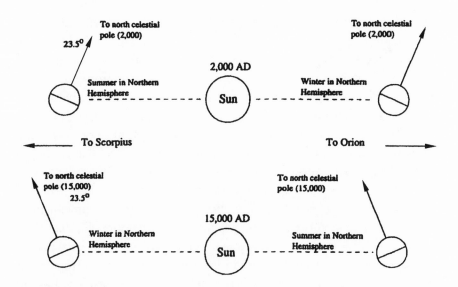

FIGURE 4.3   The orientation of the Earth's axis today and 13,000 years
from now.

above. The reason for the difference lies in the superposition of its true period with two longer periods of variations in the Earth's orbit.

The rotation of the elliptical orbit of the Earth shifts the precession from
the true period of 26,000 years to a shorter one. Perturbations from other
planets cause the orbits of all objects to rotate. Every orbit has a perihelion
point and an aphelion point (where the planet is closest and farthest from
the Sun); the line connecting the two points must pass through the Sun
and is also the longest diameter of the ellipse, its major axis. In the case of
the Earth, this "line of apsides" rotates in a backwards, or westward, direction with a period of nearly 100,000 years. Alignment of the perihelion
point with one of the solstices happens every 21,000 to 22,000 years; therefore, the time from today's extremes in the Southern Hemisphere until
they line up for the Northern Hemisphere is about half this amount, or
11,000 years.

The angle of the inclination of the earth's axis also varies slightly over
the millennia. The present value is very near 23.5 degrees, but in 2000 B.C.,
when Stonehenge was completed, it amounted to a full 24 degrees. The
heel stone at Stonehenge, over which the Sun appears to rise on June 21 as
viewed from the center of the monument, was set for that earlier and
larger value. The angle of the tilt oscillates from about 21 degrees to over

24 degrees and back in 41,000 years. This long-term motion, when combined with the precession, causes the location of the north pole in the sky and the succession of pole stars to vary slightly. Polaris was near the pole 26,000 years ago, and will mark it again that many years from now, but it was and will be at slightly different distances from the true pole because of the variations in the obliquity of the axis.

Another astronomical influence on climate arises from variations in the amount of ellipticity of the orbit itself, an even slower effect that is not periodic. The present annual variation in the distance between Earth and Sun is now about 1.6 percent, but it will change unevenly from nearly zero to some 5 percent over intervals of about 100,000 years. Summer-winter differences may be more extreme when the orbit is more elliptical than they will when it is almost circular, although precession will vary the effect. Cooler summers are thought to have a greater impact in regulating climate than have colder winters. When the Northern Hemisphere has the least insolation in the summertime, the accumulated snow of the previous winter may not melt completely but form a base upon which a new layer of snow can build. Ice ages may begin in this way.

These three slow variations—the precession, the variation in the tilt or inclination of the Earth's axis, and the change in the eccentricity of its orbit—taken together have the potential to exert a great influence on climatic conditions and on life. The Yugoslavian astronomer Milutin Milankovic (1879–1958) was the first to study these long-term effects. He suspected that these slow variations are partly responsible for the cold periods of intense glaciation we call the ice ages, a theory later confirmed by the study of ice-core samples, to be discussed later.

Other planets have their own seasons. Curiously, all the planets except Jupiter and Uranus have axial inclinations about the same as ours—about 20 to 30 degrees. Jupiter shows one extreme with a tilt of only 3 degrees, and Uranus exemplifies the other at 81 degrees. If the seasons on Jupiter were dependent upon axial tilt alone, they would be nonexistent; Uranus, with its steep tilt, would have seasons in the extreme. But, as we will see, Jupiter does have seasons; only if there were no other astronomical influence—the orbital eccentricity—Jupiter would have no seasons at all. Our figure of 1.6 percent for the Earth's orbital eccentricity gives us the third most nearly circular orbit among the planets; only Venus and Neptune have rounder orbits. The eccentricities of the orbits of Jupiter, Saturn, and Uranus are near 5 percent; those of Mars and Mercury, about 9 percent

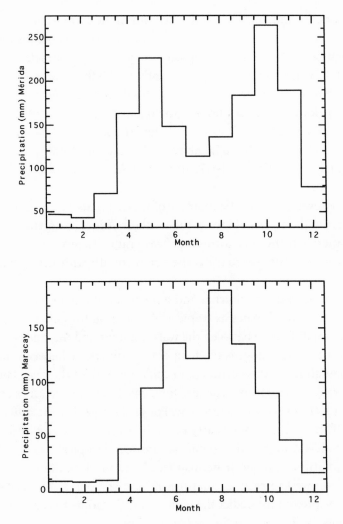

FIGURE 4.4   The annual distribution of rainfall at Mérida and Maracay, Venezuela.

and 20 percent respectively; and Pluto leads with 25 percent. Unlike the Earth's hemispheres, Jupiter's two hemispheres would be in phase; that is, both would experience summer and winter together when the giant planet is near its perihelion and aphelion points.

In the Tropics, the seasonal variation of the angle of incidence of the solar radiation is small and scarcely any major seasonal temperature variation occurs. Instead, one may find variations in the amount of water vapor in the air and in the precipitation, which even over a short distance can show significant differences. A typical example from Venezuela is given in Figure 4.4, which shows the mean monthly rainfall for Maracay and for Mérida, two cities lying only about 400 kilometers (250 miles) apart. Maracay is located near sea level in the plains near the foot of the coastal mountain range; Mérida is located in the middle of the Andes at an altitude of 1,600 meters (about 5,000 feet). Maracay shows only a single maximum in its yearly rain distribution, but Mérida exhibits a bimodal distribution.

Another even more drastic example of this nature is also found in South America. On the eastern slopes of the southern Andes in Argentina, February is battered by thunderstorms and heavy rainfalls; just across the Andes on the Chilean side, February is the driest month with nearly cloudless skies.

A comparison of the diurnal and annual variations in temperature is one way to define climates as tropical. The diurnal (or daily) range is the temperature difference between the warmest hour and the coldest hour of the day; the annual range is the temperature difference between the average temperatures of the warmest and coldest month of the year. The diurnal difference ranges from near zero for a day of heavy overcast to as much as 20°C (36°F) over a dry, sunny day. For an average day, the difference is near half this value. If the yearly range between the hottest and coldest months is less than that, the climate is considered tropical.

At midlatitudes in both hemispheres, the predominant air motion moves from west to east; for that reason farmers look at the western sky when they attempt to predict the weather for the hours to come. This feature of meteorological conditions is a consequence of the temperature difference, and therefore pressure difference, between the Tropics and each of the polar regions, plus the rotation of the Earth. At the same time, low-pressure systems rotate counterclockwise (as seen from above) in the Northern Hemisphere and clockwise in the Southern Hemisphere. Around high-pressure areas, the directions reverse in both hemispheres. (See the appendix for details.)

Note that Venus rotates comparatively slowly, one complete turn every 243 days; thus it is the object with the slowest rotation in the solar system. Because of Venus's heavy and permanent cloud cover, little if any temperature variation with latitude is expected. As a consequence of these two features, and due to its perpetual layer of clouds, this planet does not show prevailing wind directions and systematic spinning differences between the two hemispheres.

As we will see in the next chapter, comparing planets in our solar system sheds light on the unique weather patterns here on the Earth.

5

Other Worlds: Lessons from Comparative Planetology

The first day or so, we all pointed to our countries. The third or fourth day, we were pointing to our continents. By the fifth day, we were aware of only one Earth.

Prince Sultan Bin Salmon Al-Saud,
Saudi Arabian Astronaut

PLANETS ARE LIKE LANGUAGES: It is impossible to become familiar with another world, as with another tongue, without learning a lot more about one's own. The space programs that made planet exploration possible continue to reveal new facets of this planet, fully as well as any other, through a developing science known as *comparative planetology.*

The Earth is one of four planets in the inner part of the solar system, the region closest to the Sun. With Mercury, Venus, and Mars, it forms the group of terrestrial planets, so called because they share a number of common physical properties that are not found in the other planets.

The Moon can be considered a fifth member of the group, a fifth planet in a sense, because it is large for a satellite and because it shares the characteristics that identify the terrestrial group. The Moon is one of the seven large satellites in the solar system, all larger than Pluto, the smallest planet, and two of Jupiter's large four are larger than Mercury. If the Moon orbited the Sun directly, instead of the Earth, it would properly be considered a planet.

The four terrestrial planets (or the five with the Moon) are of reasonably similar size and mass when compared to the four major or giant planets, Jupiter, Saturn, Uranus, and Neptune. Each terrestrial planet has a dense, largely iron core encased in a solid mantle, which is in turn surrounded by an atmosphere of relatively small vertical extension, or only a trace of one in the cases of the Moon and Mercury.

Surely it is true that these five worlds formed from the same primordial material from which the Sun and the major planets originated, and at about the same time, some 4.6 billion years ago. All condensed from interstellar material at that time, which (like all other stars and nebulae) consisted mostly of hydrogen, the lightest and simplest element, and to a lesser degree, helium, the second lightest. More than 97 percent of all material in the universe consists of these two elements only, and this pattern is repeated in the Sun and the large planets of the outer solar system.

Our world and its terrestrial neighbors are the odd ones out: They are almost completely composed of heavier elements such as carbon, nitrogen, oxygen, silicon, aluminum, and iron, with very little hydrogen and helium. The reason for this chemical anomaly appears to lie in their relative proximity to the Sun and their relatively low gravitation. At the time of their formation, the heat from the nearby Sun drove off most of the two lightest and most volatile elements, hydrogen and helium, leaving behind the small dense cores made of heavier stuff. Unlike the major planets, which were able to retain all their original material, including their thick, dense original atmospheres dominated by hydrogen, the planets of the inner solar system condensed into small, rocky, airless worlds with mostly iron cores.

But they didn't stay airless. They had not yet cooled to their present temperatures when the considerable seismic and volcanic activity still within them spewed out the other heavier gases they had retained in a process called *outgassing*. These new atmospheres are rightly called *secondary atmospheres*. Their gases, formed of heavier atoms and molecules, move more slowly and sluggishly under identical conditions than the more buoyant hydrogen and helium. Gases such as nitrogen, oxygen, and carbon dioxide were retained by Venus, Earth, and Mars, but only marginally by the Moon and Mercury.

It is natural to expect further similarities between the constitutions of the terrestrial planets beyond those described here. Yet an examination of the composition of their present atmospheres reveals great differences between them. It is not surprising that the Moon and Mercury have scarcely any atmospheres at all. Mercury is the closest planet to the Sun; its high surface temperatures and relatively low mass have not permitted it to retain more than a trace of an atmosphere. The Moon is virtually airless for the same reason; it is farther from the Sun than Mercury, but its mass is smaller than Mercury's. If the Moon and Mercury were much farther from the Sun than they are, they would be colder and able to retain a number of gases in their atmospheres.

The importance of atmosphere and clouds in the albedo of a planet can most easily be appreciated through an examination of the atmospheres of these terrestrial planets of the inner solar system. We have defined albedo as the percentage of total incident sunlight reflected directly back into space. Venus, shrouded by clouds, has a very thick atmosphere. Mars has a thin atmosphere in which clouds are not commonly found. The albedo of Venus is over 70 percent, that of Mars only 15 percent. The Earth is in between, being partly covered with clouds, with an albedo also in between, near 35 percent. The bleak, nearly airless surfaces of the Moon and Mercury reflect only 7 to 10 percent of their sunlight. In Table 4.1, we encountered the albedos of common substances of our everyday experiences. On a sunny day, a beach or a field of snow produces much more glare than a lawn or a dirt field. Planets with ice caps and clouds reflect more than planets with only barren rock.

The ordering of these inner planets and the Moon in declining thickness and cloudiness of atmosphere from Venus to the Earth to Mars to Mercury and the Moon is not surprisingly one of declining albedo as well, as the albedos of clouds are all very high, irrespective of their chemical composition. Clouds reflect light much better than any surface material except snow and ice, and because clouds are the dominating factor here, it is evident that clouds exert a great influence on the heat of a planetary surface. We will have more to say about these planetary neighbors and their climates, if only to show that comparative planetology has taught us much of importance about our own planet.

The three largest planets of the terrestrial group have substantial atmospheres, but their surface air pressures are wildly different. The surface air pressure on Venus is about 90 times that of the Earth. In fact, Venus's thick mantle of air is about as dense as ours would be if all of our oceans boiled off into steam. Mars has a pressure less than one hundredth of ours, about 0.7 percent. Still, Mars has enough air for wind erosion and clouds to annoy Martian astronomers, if any.

Furthermore, carbon dioxide constitutes about 95 percent of the atmospheres of both Venus and Mars, whereas it accounts for only a few hundredths of 1 percent of the atmosphere of the Earth. Note also the extent of free oxygen in the Earth's atmosphere (21 percent), the second most abundant element after nitrogen (78 percent). Both nitrogen and oxygen are nearly absent in the atmospheres of the other two planets.

What happened? Why is our atmosphere at such variance with theirs? We have come to realize that originally the Earth, too, had an atmosphere dominated by carbon dioxide. But then, quite early in our planet's history, the self-replicating stuff called life formed. Life was simple at first, consisting of one-celled bacteria and algae, but even these tiny life forms could use solar energy to break down carbon dioxide and water to release oxygen through photosynthesis. Plants still use photosynthesis to enrich our air with oxygen. Absence of oxygen on our neighboring worlds implies that life also is absent there. This situation is very likely for Venus, with carbon dioxide causing an atmosphere with a runaway greenhouse effect. Venus's surface temperatures are formidably hot at some 700°F (400°C). This, with a carbon dioxide atmosphere ninety times our own, makes Venus slightly less inhabitable than Filene's Basement and like places full of bargain hunters. Human life, or at least those who shop more than we do (a category that includes most of the human race) may yet exist someday in fearfully hot, carbon dioxide-laden conditions.

Mars is a different story; its thin air does not impose such horrors to life, and it may yet exist there, but only as microbes and similar simple forms. Our globe is thus unique, at least among the objects we know, being mostly covered by oceans, having abundant free oxygen in its atmosphere, and as far as we know, being the only source of life in our solar system.

The relative surface pressures and abundances of the principal constituent gases are given in Table 5.1 for each planet. A similarity between the atmospheres of Venus and Mars is immediately apparent, whereas that of the Earth appears to be completely different, despite its intermediate position in overall air pressure and in its distance from the Sun. The difference in the absence of carbon dioxide in our atmosphere compared to the atmospheres of Venus and Mars is also apparent.

If we could decompose the carbon dioxide in the atmospheres of Venus and Mars into its constituent elements of carbon and oxygen, we would find that those planets would also possess and maintain oxygen-rich atmospheres. The main difference between the Earth's atmosphere and

Table 5.1   The atmospheres of Venus, Earth, and Mars

Planet	Surface Pressure	Major Constituent Gases
Venus	90,000 millibars	96% CO_2, 3% N_2, 0.1% H_2O
Earth	1,000 millibars	78% N_2, 21% O_2, 1.0% Ar
Mars	7 Millibars	95% CO_2, 3% N_2, 1.6% Ar

theirs can be traced to the underabundance of carbon here. The factor of more than 1,000 between the relative abundance of terrestrial atmospheric carbon and those of the other planets cannot be attributed to original formation because all three worlds were formed at the same time from the same original material. The physical conditions during the process of formation must have been quite similar on all three planets. Thus our hypothesis states that all had similar atmospheres to start with, rich in nitrogen and carbon dioxide, with no free oxygen, as is still the case on Venus and Mars.

On all terrestrial worlds, carbon and oxygen are among the lighter elements that would have been concentrated near the surfaces, whereas the heavier elements such as silicon and iron would gravitate towards their cores. Yet carbon and oxygen are too heavy to escape readily into space, unlike the lighter hydrogen and helium. Their abundances must still be nearly the same as they were when the planets had cooled enough to form their permanent solid crusts, and above the crusts the secondary atmospheres had seeped out of igneous rock at temperatures low enough for water to exist in liquid form. Because oxygen reacts readily with many other elements, it is unlikely to have existed in free form. It must have formed many oxides, including carbon dioxide.

Considerable evidence (including geological evidence) supports the hypothesis that at first the secondary atmospheres of all of these planets contained little or no free oxygen. In the oldest formations on the surface of the Earth, pebbles have been found whose form indicates exposure to the polishing action of long-time running water; this implies that the pebbles have long been in fast-running creeks, where they were exposed not only to the water but also to the air above. These pebbles consist of a rapidly oxidizing substance (uranium pitchblende), yet they show no evidence of oxidization. No free oxygen has been in the air during their lifetimes in the riverbeds. Where, then, did the carbon or carbon dioxide go, if carbon was originally among the most abundant components of the Earth's atmosphere? Carbon is too heavy to have escaped into space; it should still be present in amounts not much below the original amounts, and we must explain how we know where it is and how it got there, and finally, where the free oxygen came from. In all these questions, the primary ingredient consists of organic processes; that is, life, even in its most primitive form.

The original secondary atmospheres formed by outgassing over the surfaces of the terrestrial planets must have consisted of carbon dioxide, water vapor, nitrogen, argon, and a few trace gases. Weather and the seasons as we understand them today were already present. Heat, cold, wind, and rain were already acting on the surface of a lifeless Earth and eroding it for nearly 1 billion years. Creeks and rivers, flowing through mountains and plains, dissolved on their way whatever soluble substances they encountered and carried them to the oceans; the oceans originally consisted of fresh water, but through this action they became more and more saline. This scenario eventually led to the development of the first forms of life. Still no one agrees about the birthplace of life. Whether life first formed on the dry surface or in the oceans, organic life was and still is based on photochemical processes; these decompose carbon dioxide, use the carbon for their structures, and leave oxygen free to escape into the air. Later on, when oxygen had already reached a significant abundance, nature devised new processes that use oxygen as well. However, even when dense vegetation completely covered the Earth and algae in the oceans grew abundant, the total loss of carbon dioxide from the air was still not significant.

If carbon dioxide is present in the air, it is also present in the oceans, at least in the layers near the surface. This gas can penetrate ocean water and remain in it much as it does in a soda bottle; it combines with the water to make an acid, but some of it also escapes readily back into the air. A certain equilibrium results and depends on two factors: the pressure of the gaseous carbon dioxide and the water temperature. Thus the amount of carbon dioxide in the sea depends upon the amount present in the atmosphere. The sea temperature is also a factor; warm water cannot contain as much carbon dioxide as cool water. The carbon dioxide in a soda bottle presents a similar example; when it is heated, some of the carbon dioxide escapes. Scientists estimate that at present, some sixty times as much carbon dioxide exists in the oceans as in the air and that marine life depends entirely upon the carbon dioxide in the water.

Altogether the air, the sea, and vegetation, dead or alive, account for only a few percent of the original carbon. It is evident that there are other places in which nature has stored the missing carbon, and indeed there are several. On dry land, carbon dioxide is continuously recycled from the air into the vegetation, and from there into animals, and then by decomposition into the soil. From the soil, it can pass back into the air to complete the cycle.

Another equally important cyclic process happens in the water. This process is different to that of the air. Water extracts calcium carbonates, for which no equivalent is found in the air. Many forms of marine life have shells made of calcium carbonate. The shells contain carbon dioxide and are insoluble in sea water, although some organic forms die, decompose, and are returned to the gaseous phase as they are in the air. When a shell-bearing animal dies, its shell sinks to the ocean floor and remains there. In this manner, thick layers of calcium carbonate have accumulated over millions of years. Tectonic motions of the plates that form the Earth's crust eventually bring some of the layers to the surface, converting them into dry land. Many of our present-day mountain ranges once lay on an ocean floor. The motions form just one of the ways by which carbon dioxide can be removed from the atmosphere and stored permanently as a solid (coal, carbonates, limestone) or liquid (oil) on or below the surface. The large deposits of oil, coal, and natural gas are examples of carbon removal and its sequestration.

Combined with water vapor in the air, the atmospheric carbon dioxide forms carbonic acid. Carbonic acid is an effective weathering agent, acting on silicate rock. Rain washes away the reaction product; it eventually winds up in lakes or the ocean, where it will form limestone. Such deposits are mostly found in shallow off-shore waters. Limestone contains carbon and is not water soluble; here is another case in nature that leads to a permanent carbon storage.

How interesting it would be if we could reconstruct the history of carbon dioxide abundance in the Earth's atmosphere from the very beginning, more than 4 billion years ago. Later on, we will find a few samples of air that are several hundred thousand years old, and we now have indirect methods that estimate carbon abundance at earlier times. If all the carbon dioxide and carbon deposits were removed from the ground and restored to the air, and the pure carbon were to recombine with the oxygen, we would find an atmosphere whose constituency would be little different from the atmospheres of Venus and Mars. Carbon dioxide would be the most plentiful gas, followed by nitrogen. Evidently, organic life has played a decisive role in the atmosphere of our Earth. It is ironic that life has managed to eliminate most of the atmospheric carbon dioxide, and in doing so, has threatened its own existence.

Every world but ours presents hostile conditions for our sustenance. The possibility of terraforming the Moon or Mars, of making conditions on their surfaces much more Earthlike, has deservedly received attention; but it is clear that the necessary technology lies decades or centuries in the future, if it is developed at all. Until that hypothetical time, all 6 billion of us must live here on this one globe. Let's not spoil it. As Harrell Graham has said, "Good planets are hard to find."

6

WEATHER
WISDOM AND LORE

The Moon and the weather may change together;
But change of the Moon does not change the weather.
If we'd no Moon at all, and that would seem strange,
We still should have weather that's subject to change.

ARTHUR MACHEN, *NOTES AND QUERIES*, 1882

PEOPLE LEARNED TO INTERPRET SIGNS in the sky for coming weather long before they learned to chronicle historical events through a written record. Some predictions belonged to the realm of fancy and superstition, but others came from experience in that they were correct more often than not.

Many of the omens, whether or not grounded in reality, came about from devout and transcendent beliefs, but others arose from observation, sometimes mistaken observation but commonly rational. We can classify these auguries according to their observed phenomena; of the largest groups, one has to do with the appearance of the Sun and the Moon, and another involves the presence, appearance, and color of clouds and clear sky. Other groups are based on animal behavior, on wind and precipitation in all its forms, on weather on specific holidays or other dates, on seasons during the year, on aches and pains, and, in recent times, the weather as noted by such instruments as the barometer.

The Moon has played a large role in weather lore ever since people have contemplated the possible association between the two. Today, with hindsight, we can divide this association into two groups. One is the orientation of the Moon in the sky and the other consists of changes in the Moon's appearance imposed by the Earth's atmosphere.

In many places on the Earth, the wet and dry seasons or the hot and cold seasons coincide with extremes in the orientation of the ecliptic in the sky and consequently of the Moon's appearance. But no causal relationship has been established, and it is unlikely that one exists. The horns of the waxing crescent Moon in the evening sky after sunset can point up or sideways, depending on one's latitude and the time of year. If the times happen to correlate with seasonal climatic effects such as a rainy or dry season, they confirm Machen's first line but do not create a causal relationship between the Moon and weather.

The belief that the orientation of the horns or points of the crescent Moon to the horizon affected rainfall points to the happenstance that a rainy season coincides with the inclination of the Moon's path through the sky in some locations; it cannot hold over the entire globe.

Another belief, almost universal, holds that the weather changes with the phases of the Moon. When questioned about it, believers admit that the weather change could happen at any of the four phases, and that the change may occur up to three days earlier or later than a given phase. This takes care of the entire lunar cycle.

Even among those signs with some credence, however, a danger lurks. The words "always" and "never" are understandably rare in the behavior of weather, as they are in human behavior, and for much the same reason. Too many unpredictable variables exist in both cases for an ever-reliable rule. Yet patterns emerge that can be useful guides if they are not taken as absolutes. Cyclic phenomena are particularly exasperating; as soon as one phenomenon is thought to depend on another, an exception comes along to spoil it. As soon as the division or league with the winning team of the World Series or the Superbowl appears to foretell the stock market or the outcome of an election, it breaks down. One example, much discussed during the Truman administration, was that every one of the nine American presidents of this century up to that time had double letters occurring somewhere in his name; some people gave this phenomenon credence as a rule. But only two of the nine presidents since that time, Kennedy and Clinton, have followed the pattern; the "rule" has long been abandoned.

Even better known is the happenstance, considered a reliable jinx by many, of the death in office of those presidents elected in years divisible by twenty; but Ronald Reagan spoiled the rule in 1989 when he left office very much alive. Promoters of the "jinx" often conveniently forget that the first two presidents to qualify, Jefferson and Monroe, left office alive, and Zachary Taylor died in office after being elected in 1848.

If we keep in mind that the better weather lore sayings consist of rules with exceptions, not of axioms, they can be useful.

•   •   •

In our midlatitudes, most weather arrives at a specific location from the west, or nearly so. It may arrive from the northwest at times, and from the southwest at others, but only rarely does it move in from any of the other points of the compass.

The reasons for this predominance are straightforward. Remember that cold air is heavier and denser than warm air; as such it seeks to compact itself more tightly to the ground. At a given altitude, a mass of cold air registers a lower air pressure aloft than does a warmer mass; when the air is cold more of it is compacted near the surface than would be the case with warm air. We can think of the Northern (or the Southern) Hemisphere as a tripartite model; that is, each can be loosely divided into three separate latitude zones.

The zone extending northward from the equator to about 30 degrees north or south latitude is roughly equivalent to the Tropics, where the climate is warm and summery throughout the year. The arctic, or polar, region, the area poleward of about 60 degrees latitude, north and south, is perpetually wintry and cold with only occasional intrusions of mild air. The temperate midlatitudes have marked seasons, with mostly tropical conditions in the summertime and often subfreezing conditions in the wintertime. But in any season a temperature gradient from tropical to near-polar areas occurs somewhere in the midlatitudes, moving northward and back southward with the Sun.

A gradient in temperature gives rise to one in pressure at high elevations. The colder air to the north packs down more than the warmer. Because air tends to fill a vacuum, and also because air flows much more freely horizontally than vertically, a pressure gradient wind arises and seeks to blow towards the poles and away from the equator in both hemispheres. The equator would be hotter and the poles colder than they are if no heat were transferred from the Tropics to the polar regions.

If the Earth did not rotate, the picture would be simple. Air near the Equator heats up and rises. Having nowhere else to go, the heated air flows poleward at high altitudes. To compensate for the absence of air near the surface, cooler air moves more gently toward the Equator near the surface and tends to sink near the polar regions. This action sets up a hemispheric heat engine, not unlike the onshore and offshore breezes set up near coastlines (to be discussed in Chapter 9). On the global scale, this simple model is referred to as a Hadley cell, named after the eighteenth-century English scientist who first modeled it (see Figure 6.1).

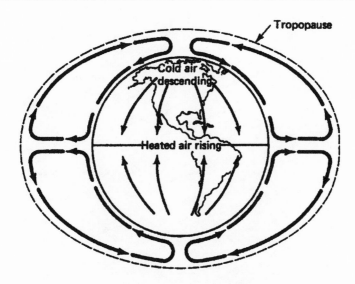

FIGURE 6.1   Global air circulation on a nonrotating Earth.
This is the well-known Hadley cell model.

But the Earth rotates, complicating the picture to a great extent. Because the Earth spins around once every twenty-four hours, and its circumference is about 25,000 miles (about 40,000 kilometers), we spin right along with it at a speed of 25,000/24, or just over 1,000 miles (1,600 kilometers) per hour. But this speed decreases with distance from the equator; closer to the pole, it is slower (see Figure 6.2), and at the pole, it is zero. In twenty-four hours at the pole, one spins around completely without moving from a given spot. Santa Claus's house just rotates on itself, with every window facing south all the time.

An epiphany of sorts occurred when scientists looked at Venus for the first time from a nearby space probe. Because Venus is perpetually cloudy, scientists knew that it rotated slowly, but the lack of visible surface features prevented a better determination from the Earth. The biggest surprise was that Venus rotates much more sluggishly than we had imagined, and it takes about 243 of our days for it to rotate just once! Venus has the longest "day" of any known object, and with a diameter just smaller than that of the Earth, its rotation speed at its equator is only 4 miles (6 kilometers) per hour, the pace of a brisk walk. Venus has so little rotation that it shows a Hadley cell model of air circulation. There before us was the model first proposed for our spinning planet over two centuries ago. In Figure 6.3, both planets appear shown to scale, Venus being the slightly smaller

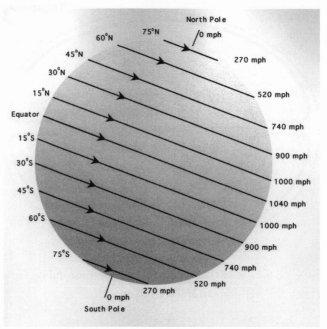

FIGURE 6.2   Rotational speed as a function of latitude
on the Earth's surface.

planet. Note that the Earth is replete with the vortices and swirls characteristic of a spinning world, but Venus is layered and has almost no turbulence in its atmosphere.

•   •   •

Here, we must pause to explain the *Coriolis Effect,* the name for an apparent curve in the motion of a projectile, missile, or as here, wind. If a ship in
the Northern Hemisphere fires a missile at a target ship to the north (at a
higher latitude), the missile would deviate to the east of the target ship because of the Earth's easterly rotation. But by knowing the latitudes of the
two ships and the speed of the missile, an artillery man can calculate a cor-
rection for the motion, known as the Coriolis Effect, which occurs everywhere on the Earth except along the equator. The deviation is to the right
north of the equator and to the left south of that line.

The Coriolis Effect is also responsible for the deviation of the pressure
gradient wind in the midlatitudes from its otherwise northerly (or
southerly) direction to the east in both hemispheres, thus giving us our familiar prevailing westerly winds, or westerlies. A balance is created between the two effects (pressure gradient wind and Coriolis Effect); this

FIGURE 6.3   The sizes of Earth and Venus to scale, showing cloud patterns on Venus and the more turbulent Earth (NASA).

balance is called the *geostrophic balance,* which produces the *geostrophic wind.* Moderate deviations occur to disturb the wind from a straight westerly direction, a little to the north or south, but rarely so much as to shift the direction far from directly west.

With this in mind, let's look at one of the most widely known weather legends of all:

> When it is evening, ye say, it will be fair weather: for
> the sky is red. And in the morning, it will be foul weather
> today: for the sky is red and lowering.
> O ye hypocrites, ye can discern the face of the sky; but
> can ye not discern the signs of the times?
> **(Jesus to the Pharisees, Matthew 16:2,3)**

This quotation from the Bible is found in many variations. Probably the best known version is this:

> Red sky in morning, sailors take warning.
> Red sky at night, sailors' delight.
> **Old European chant**

A more extensive variant goes as follows:

Evening red and morning gray
Sets the traveler on his way;
Evening gray and morning red,
Brings down rain upon his head.

All these pertain to the general circulation model posed above: In our midlatitude regions, weather systems move from the west to the east. Sky conditions to the west, to the part of the heavens most brightly illuminated by the setting Sun, are indicative of the weather to come, whereas those to our east, shining in the dawn sky, correspond to weather that has gone by. The gist of these three versions applies directly to that realization and may well have been perceived before civilizations began.

The chances are, as the ancients of many lands came to realize, that a red sky to the east (thus in the morning light) means that fair weather has gone by, the humidity is increasing, and foul weather may come. To the west, a red sky predicts the opposite, that fair weather is on its way.

An evening sky gray in the west means that clouds heavy enough for rain are coming, but if the sky is gray to the east, the clouds would presumably be fleeing away. These omens are more often right than wrong, and thus qualify as legitimate, if not highly accurate, weather predictors. Similarly, rainbows foreordain wet and dry conditions much as do gray skies. A rainbow to the west, or more exactly to the windward, means that rain (droplets in the air) is the likely cause for the rainbow's visibility. A rainbow to the east, to leeward, means the rain is moving away. This feature is illustrated in Figure 6.4, in which windward is associated with a westerly direction from which most wind and weather comes, and leeward is to the east. Thus:

Rainbow to windward, foul fairs the day,
Rainbow to leeward, damp runs away.

•   •   •

Small showers last long,
but sudden storms are short.
William Shakespeare, *Richard III*

This couplet addresses another well-known weather phenomenon, the passage of fronts. Most fronts are of two kinds, warm and cold, and each represents a zone between two masses of air with contrasting properties.

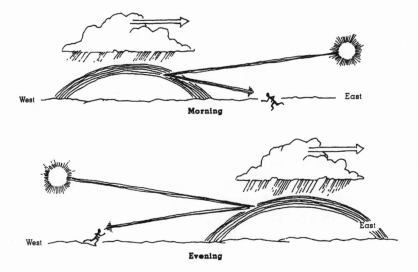

FIGURE 6.4   Rainbows in morning and evening; in both cases they appear in the opposite direction to the Sun.

Think of air masses and their frontal boundaries in three dimensions. The front is actually a plane or curved surface, perpendicular or tilted with respect to the ground, which separates one mass from its neighbor. Where the plane intersects the surface, the line is drawn on weather maps and called a *front*. As air masses glide by from west to east, the one with the colder air wedges under the other. A warm front passes when we see the warmer air riding up and over the colder. The cooler air lags behind, and a wide frontal zone is usually present.

Fronts between air masses are sources of turbulent weather; along fronts, air friction along the border between two air masses moving in different directions causes turbulent air motion, and precipitation develops. Because of the length of the warm front, and because it passes a point on the surface at perhaps 10 to 20 miles per hour (16 to 32 kilometers per hour), the rain can be a drawn-out affair, lasting a day or longer.

The cold front is sharper, narrower, produces harsher weather, and moves twice as fast past a given point. Here the oncoming cold air wedges under the warmer air and forces it aloft. The warm air cools, quite suddenly at times, and relieves itself of its moisture. Heavy rain, wind squalls, and thunderstorms are frequent along the front, particularly if the temperature difference between the masses is great.

May produces the most violent local storms, thunderstorms, and at worst, tornadoes. May's weather brings summertime heat and moisture content, but it can also bear the chill of winter, or at least early spring. By July and August, all weather is hot, and air masses may differ only in water content. The virulence is not so great as it was earlier in the year because the displacement of one air mass by another of very similar properties causes less unrest. The May cold front can drop the temperature by 30°F or more, but later on in the season a lowering of more than 10 degrees is uncommon. Storms that form locally and not along fronts, however, are most likely to form in the moist, nearly stagnant air so common to summer months. Similarly, near the equator, fronts are rare because all air masses are hot, and they do not usually vary much in moisture content. Fronts mark the boundaries of definable air masses and therefore aren't much in evidence in the Tropics.

No matter the season, warm front precipitation is gentle and lasts for a day or more; that of a cold front may pass in only a few hours, but with greater amounts. Thus:

Rain long foretold, long last;
Short notice, soon past.

Another well-known bit of weather lore:

Dry grass at morning light
Look for rain before the night,
When the dew is on the grass,
Rain will never come to pass.

Here again, the truth of this lies in two points mentioned earlier. Land cools off at night much more rapidly than does water, and furthermore, air near the surface quickly takes on the properties of that surface. We said in Chapter 2 that relatively dry air has a dew-point temperature well below its own temperature and must cool dramatically to reach the dew point, where saturation occurs. Moist air has a dew point just below the air temperature, thus condensation is much more likely to take place.

On a crisp, clear evening, the ground and the air just above it cool quickly from their high temperatures earlier in the sunny afternoon. The air nearest the surface cools the fastest and deposits its water vapor on the ground. Blades of grass have much more surface area per square foot or meter of level surface than does flat ground and they hold more droplets

FIGURE 6.5   The Wallensee, a lake in Switzerland.
Clouds frequently form along mountain summits, as is shown here.

as a result. Under cloudy or hazy conditions, the surface cools off comparatively little, partly because it heated up less during the daytime and partly because clouds promote the greenhouse effect. Night fogs can develop in the same way; if a night is clear and windless, a low ground fog may develop. Sometimes, the fog remains so close to the turf that the horizontal visibility can be a city block or less, yet the brighter stars are clearly visible overhead.

> When old man mountain wears his hat,
> Ye men of the valley beware of that.

Whenever air blowing along fairly level ground or the sea encounters a mountain or a range of mountains, it must rise to get over or around them. This rising air, known as *orographic lifting*, invariably cools as it ascends. If the air rises high enough, it may reach saturation. In this event, clouds form from the water droplets as they condense. Fog is cloud in contact with the ground; cloud appears as fog, especially to a person surrounded by it. Clouds commonly form along the sides or at the summits of hills and mountains (see Figure 6.5). If a cloud forms there, if old man mountain wears a hat, and particularly if the cloud continues to grow, moisture content is rising, the increase frequently leading to rain or snow.

The calendar provides occasions for many adages about the weather and, like those quoted above, most are more often true than false. The lag of the seasons has become a source for any number of old saws. Temperature extremes for the year do not occur with the maximum and minimum in insolation, which take place at the solstices on or about June 21 and December 22. Summer weather is hottest in July and August and the coldest days occur in January and February, earlier in continental climates and later in the maritime regions dominated by the oceans. The thermal inertia of the atmosphere causes this delay, even though the Sun has started its ascent and descent from the winter and summer solstices. The following proverb speaks directly to this phenomenon:

When the days begin to lengthen,
The cold begins to strengthen.

•   •   •

If it rains on Saint Swithin's Day,
It will rain for forty days straight.

Saint Swithin's Day (July 15) is one of the better-known holidays or special days associated with weather. There are many such days; one well-known example falls on February 2, known as Candlemas Day in Britain and Groundhog Day in the United States. In both cases, custom has it that a clear sunny day leads to a late spring, as in Punxsutawney, Pennsylvania, where a groundhog may see his shadow, indicative of a long winter, or he may not. Along the same lines:

The first three days of a season
Rule the weather for that season.

These proverbs, and many others like them, suggest that future weather is determined by that of a particular day or days. This does not hold up to inspection. Unlike the adages we have discussed, these proverbs do not come true more often than randomness would predict. Rarely is a specific period associated more than randomly with a specific weather condition. One of the possible exceptions is Dogwood Winter. In his delightful book, *Our American Weather*, George H. T. Kimble writes this: "One of the things that never ceases to amaze the student of North American weather is its errancy. Try to make rules for it, and you will find that they are more often honored in the breach than in the observance."

In this vein, a nineteenth-century meteorologist, Alexander Buchan, found among weather records for Scotland a smooth rise and decline in temperature from the coldest of winter to the hottest of summer and back again throughout the year; but there were anomalies, departures from the temperature curve, during certain times. These have purportedly been confirmed by observers in the United States. The well-known January thaw appeared to be followed by a very cold period in mid-February, and a markedly warm spell seemed to occur during the first half of December.

But the most persistent legend has to do with the second week of May, during which a temperature relapse takes place, usually accompanied by rainy and windy weather. This short return to near-winter conditions is called *Dogwood Winter* in eastern North America, occurring as it does when the magnificent dogwood tree is in bloom. That this phenomenon falls during that second week of May, and not the first or third week, or not at all, is moot. On both sides of the Atlantic Ocean the tendency is noted, but in this case, the past may not guide the future. One of the authors has noted that this period may have a slight tendency toward cold, raw weather, but in almost as many years the period is as warm or warmer than average.

In Germany, June 27 is known as the *Siebenschlafer*, which means "seven-day-sleeper"; when it rains on that day, it is supposed to rain for seven weeks. May 11, 12, and 13 are the *Eisheiligen* in Germany (the Ice Saints in Britain); a return of almost winter conditions is expected on these days, just as with Dogwood Winter in the United States. Some in Germany believe that an unusually rich harvest of chestnuts is followed by a severe winter. (This can already be predicted in the spring when the trees bloom.)

The next two sayings have more truth to them:

March in January, January in March.

and

Green Christmas, white Easter.

Both sayings predict a severe second half of a winter if the first half is mild. We find some truth to this, for weather systems that seem to stay in place for weeks on end do slowly move eastward and can take up residence half a continent or ocean away. The reverse does not happen often, and a severe first half is about as likely to be followed by severe weather as by milder.

If the sky beyond the clouds is blue,
Be glad, there is a picnic for you.

Clouds, of course, often make fine weather forecasters. Some of these are covered above in the descriptions of fronts and the appearances of the Moon and Sun. The rate at which clouds thicken from high, thin cirrus to heavy rain clouds is taken as a harbinger of frontal passages.

Clouds are of three basic types and feature ten subtypes, recognized internationally. Cumulus is the name we give to the rounded, puffy clouds that are usually present with fair weather. The word "cloud" evokes an image of the cumulus more than it does any other type of cloud. Claude Debussy must have had cumulus clouds in mind when he composed "Nuages," the first of his three orchestral nocturnes; indeed, cumulus clouds float languidly across a blue summer sky, as does Debussy's whole-tone music. The couplet above points to their primary characteristic and to their origin—that they form mostly over land and rise in *thermals,* rising blobs of air heated by their proximity to the hot ground. As the blobs of air rise and cool, they may reach condensation and a cloud of water droplets may form, typically 0.5 to 2 kilometers (0.3 mile to 1 mile) above the surface. Because sunlight is the heat source for cumulus clouds, they frequently dissipate around sunset. The blue sky beyond is a common feature, and unless cumulus congest to the point of rain, the clear sky remains to assure a starry evening.

The thin feathers and mares' tails are cirrus clouds; they are high and so are composed of ice crystals. The air at 5 to 10 kilometers (3 to 6 miles) aloft is thin and the clouds are always too frail and insubstantial to block sunlight or to generate precipitation. Stratus clouds are the third principal type; they form strata or layers, as their name implies, and appear gray and usually featureless from below. The gray cloudy day is a day of stratus clouds, seemingly without end. Illustrations of each primary type are shown in Figures 6.6, 6.7, and 6.8.

Wind and clouds, and the appearance of the sky, are authors of weather maxims, true or false. This poem expresses a widely circulated notion of the winds:

When the wind is in the north,
The skillful fisher goes not forth;
When the wind is in the east,
'Tis good for neither man nor beast;

FIGURE 6.6   An example of cumulus clouds.

When the wind is in the south,
It blows the flies in the fish's mouth;
But when the wind is in the west,
There it is the very best.
Izaak Walton, *The Compleat Angler*

A north wind (wind blowing from, not to, the north) generally means cold weather; the wind creates a wind chill and the air can be unstable at times, bringing rain or snow. From the south we get our hot, muggy air, chiefly in summer. Then the three Hs descend: hot, humid, and hazy. This air carries the most contamination and pollution and is the most un-healthy of all.

An east wind is uncommon; it runs counter to the prevailing westerlies in the temperate regions. It usually means that air is circulating around a low-pressure area, or *low*; it carries over a relatively small surface area and for a few days at the longest. *Nor'easter* is the local term used in the eastern United States and Canada for the infamous frequent winter storms that move up and just off the east coast; they account for the heaviest snowfalls in that locality, or rainfalls if the temperature remains above the freezing mark; their overall progress is from the southwest to the northeast, but the winds swirling around their centers could be from any direction.

FIGURE 6.7   An example of cirrus clouds.

A local wrinkle in the normal flow can also create a *back-door cold front,* the front passing from east to west; but usually within a day it stops and becomes a stationary front and then moves eastward as a normal warm front.

The west wind, the usually gentle west wind, is the culmination of fair weather, from the kite-flying windy days of March to Indian summer in the fall, the warm golden days that follow a first frost.

FIGURE 6.8   An example of stratus clouds.

A change in the direction of the wind provides an even better weather guide. The wind is said to be veering when it moves to the right in a clockwise direction, whereas a backing wind shifts leftward in the opposite direction. A weather or wind vane is designed to point in the direction from which the wind is coming: If the vane gradually swings from left to right along the horizon, the wind is veering; from right to left identifies a backing condition. Those who sail know the following couplet:

A veering wind, fair weather;
A backing wind, foul weather.

or this longer example:

Winds that swing against the Sun,
And winds that bring the rain are one.
Winds that swing round with the Sun,
Keep the rainstorm on the run.

The second verse tells us that in the Northern Hemisphere, the Sun, rising in the east, moves across the sky in a rightward or veering direction to its rendezvous with the western horizon at dusk. Sailboating makes use of the association between wind and rain, as in

> If the rain precedes the wind,
> Lower your topsails and take them in;
> If the wind precedes the rain,
> Lower your topsails and hoist them again.

In general, when rain precedes strong winds in the Northern Hemisphere, the tempest to come is often savage. Otherwise, it may be weak and pass quickly.

In the last century, the Dutch climatologist Buys-Ballot found a direct relation between the direction of the wind and the directions toward low or high pressure. His law states that when one stands with one's back to the wind, the right hand points toward high pressure and the left hand toward low pressure. South of the equator, these directions reverse.

This all develops from one of the universal rules in weather. North of the equator, the wind always blows clockwise around a high and counterclockwise around a low. Again, the opposite happens to the south of the equator.

Now imagine Buys-Ballot (or anyone else) standing anywhere near a high or a low. In every case, when the wind is blowing from behind him, the high pressure is to his right and the low to the left. Highs and lows are centers of clear and stormy weather in most cases; we will see later how and why this is so.

> Buys-Ballot stood upon the deck,
> His face was toward the flagpole.
> High pressure air lay to his right,
> and the wind blew up his shirtsleeves.

•   •   •

Animal behavior is yet one more source of weather lore. Almost every species is represented by at least one legend. Here is a general one:

> When waters rise, the fish eat ants,
> When waters sink, the ants eat fish.

Although this couplet is only marginally connected to the weather, in times of flood or drought it has a certain reality. This saying has more immediacy:

When cockroaches fly, rain will come.

But other sayings contend that when cows stand with their backs to the wind, or when they low, or when they lie down in the morning, or when they don't produce milk, expect rain soon. Expect it also when bears are restless, when wolves howl, when owls hoot, or when ravens crow.

In one sense, all these old maxims are true; it will rain sometime after each one of them. But that does not tell us about their ability to predict the weather within a meaningful time, nor does it suggest a causal relationship. It will also rain after every presentation of a Nobel prize to an American, or every hole-in-one scored on the golf course, or the discovery of a previously unheard composition by Vivaldi, or another Picasso retrospective exhibition, or even another statement by Federal Reserve Chairman Greenspan that causes fibrillation in the Dow. Most observations about the behavior of animals, birds, woolly caterpillars, or people have little or no value as omens.

Two other sayings, both well known in Venezuela, appear more reliable. The first states that the *grillos,* or cicadas, resembling grasshoppers by amassing by the billions, begin to chirp just before the rainy season starts. This happens most commonly in the *llanos,* or plains, and the sounds can be deafening.

A second weather observation also pertains to the few days before the rainy season when the centipedes emerge from their holes and look for shelter in homes and office buildings.

At the beginning of this chapter, we mentioned the influence of the Moon on weather lore and gave an example. Arthur Machen's epigraph notes that many of the inferences of this influence have no basis in fact.

Some apparitions of the Moon do correlate with weather changes; who has not heard that a ring around the Moon is a harbinger of foul weather? This phenomenon deserves some comment. First, two kinds of rings can surround the Moon at night or the Sun in the daytime. One is a wide ring of about 23 degrees in radius that can fill a fair portion of the sky; this type of ring is called a *halo* and is caused when moonlight passes through ice crystals, usually near the stratosphere. Such phenomena are associated with cirrus clouds, those little feathery clouds that are too thin to block light from the Sun or Moon. Cirrus clouds, especially when they thicken and are followed by thicker and heavier clouds at a lower altitude, may be associated with an approaching front bearing rain or snow. Because a halo is not always indicative of precipitation, it is an unreliable omen. Nevertheless, if the halo becomes more intense as the night wears on, it may be of some use as a guide to coming conditions. The other type of ring, called a *corona,* is brightest at the limb of the Moon (or Sun) and fades away a degree or two outward from its edge. A corona indicates a lower cloud composed of water droplets; its value in predicting inclement weather is about as good as the halo. We should note the Moon owes its unique place in weather lore primarily to its brightness in the night sky. Stars are too faint to illuminate a halo or corona, although occasionally a bright planet such as Venus or Jupiter is circled by one or the other.

Other Moon signs have been used to predict the weather:

Clear Moon, frost soon.

and

A pale Moon doth rain.

The second of these falls into the same group as coronae and halos. When the Moon pales, clouds may be on the increase. The first saying is often true because on clear nights the cooling of the Earth's surface is greatest; the drop in the temperature can cause the condensation on plants and lawns we know as frost or dew, depending on whether the temperature does or does not reach the freezing point.

Finally, we cite an old adage from Sir Patrick Spens about the appearance of the crescent Moon:

I saw the new moon late yestreen
With the old moon in its arm.
If ye be goin' to sea, sir
I fear you'll come to harm.

The *Old Moon* is another name for Earthshine caused by the double reflection of sunlight from the Earth to our west when it is still daylight, and then off the Moon and back to us. At *New Moon,* because the Earth appears almost full from the Moon, the Earth's light illuminates the Moon. The bright new, or crescent, Moon is illuminated by direct sunlight; the faint rest of the Moon, called the Old Moon, is sometimes faintly visible. No connection has been found between the visibility and brightness of this feature and the chances of a voyage encountering inclement weather.

•   •   •

Each of us has vivid memories of weather experiences; these memories often ripen into lore whether or not we share them with others; In either event, weather lore is natural and provides color to our weather experiences. One of the authors recalls a brilliant rainbow displayed across the entire city of Edinburgh as seen from Edinburgh Castle; a local shower of some intensity on one side of a residential street for some minutes before extending to the other; and a Memorial Day picnic during which a sudden 30-degree drop in temperature sent him scurrying to unpack his winter gear. Once, while driving in Michigan, his car was blown sideways into a ditch, perhaps by a minor tornado. The other author remembers equally such impressive weather as a major rain-caused landslide onto a road, after which a local farmer picked up an egg-shaped stone. When the author asked the gentleman whether another landslide might happen, he paused and answered, "Whoever laid this egg will come back for it."

STORMY WEATHER

Blow winds, and crack your cheeks! Rage, blow!
You cataracts and hurricanes, spout
Till you have drenched our steeples, drowned the cocks.

WILLIAM SHAKESPEARE, *KING LEAR*

Thus does the old king direct his prattle against the fury of the storm, personifying its elements as he and his fool roam about in it. Shakespeare is one among many writers who adapts a storm to his tragedy *King Lear*.

Europe encounters storms of the same type as those in North America. Storms have made their way into literature, as when Emily Brontë's Cathy and Heathcliff roamed the moors surrounding Heathcliff's home, Wuthering Heights (also the title of the book), and into music, when Benjamin Britten's Peter Grimes made his fateful way along the stark East Anglican coast. A better-known musical storm, depicted in the fourth movement of Beethoven's sixth symphony *(Pastoral)*, suggests a briefer interlude of rain, perhaps akin to a summer shower or thundershower. And what of that raging storm surrounding King Lear and his Fool when they were lost on the moors, a storm that rumbles its bellyful throughout much of the third act of that drama? Despite the references to thunder and lightning, this seems more than a thundershower, more of a longer blow, not unlike the famous nor'easters of the United States.

For centuries, stormy weather has held a prominent place in song. "Don't know why, there's no Sun up in that sky, stormy weather," begin the lyrics to the popular blues ballad of long ago. At almost every point on the Earth's surface, clouds, rain, snow, and other elements of inclement weather come and go with some regularity. Before the space age, we pictured our planet from space as a world of blues, greens, and browns, and some white to represent the polar regions. Usually, the artist didn't think of adding a healthy dose of clouds, but in reality clouds cover about half

FIGURE 7.1 Hurricane John, a typical hurricane showing a well-formed eye at its center.

the globe at any one time, many of them dumping precipitation. But now the world has seen NASA's stunning photos of the real Earth, and artists get it right.

Stormy weather comes in two main forms, one originating in the Tropics, and the other in the midlatitudes. Fundamental differences exist between them. Storms originating near the equator have many names, but they are basically the same kind of creature. Those spawned in the Atlantic Ocean are *hurricanes,* a word from the native West Indies. The same storms are called *typhoons* when rising from Pacific waters, and *cyclones* when originating in the Indian Ocean. Figure 7.1 shows a typical well-formed hurricane.

Storms form a few degrees north or south of the equator, just far enough for a slight Coriolis Effect (see Appendix IV) to be present. They start as tropical depressions and, if conditions are favorable, they develop into swirling tropical storms. They tend to be circular, with an *eye,* a calm area, in the center. At the edge of the eye, the winds are strongest and the rainfall is heaviest, and when the winds rise to a steady velocity of 40 miles

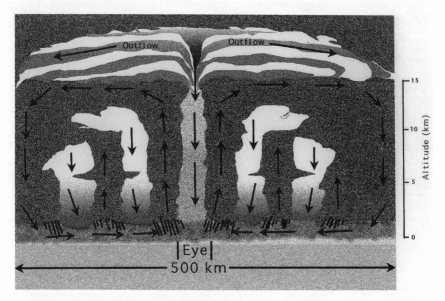

FIGURE 7.2   Vertical cross section of a hurricane, showing air motions, rain, and clouds.

per hour, they are reclassified as tropical storms and receive a name. The naming procedure is almost a half century old; at first, they were given only feminine names, but later every other storm received a masculine name (see Figure 7.2).

Tropical storms move westward at this point, and drift slightly away from the equator. Under extreme conditions they strengthen further, and if the maximum steady wind is measured at 75 miles per hour or more, they graduate to the status of hurricanes, or typhoons, or cyclones, depending on their birthplaces. The severity occurs as a result of the system's tightening into a circular pattern. Five characteristics distinguish hurricanes from their extratropical counterparts.

First, hurricanes form only during certain times of the year. In the North Atlantic Ocean, the hurricane season starts on June 1 and ends on November 30. The majority of hurricanes form between August and October, with a maximum around September 10, the time of year when the ocean waters are warmest. Hurricanes feed on water warmer than 27°C (or 80°F), and the hotter the better. Over water cooler than this, or over land, they quickly weaken and fall apart, their second feature. They are not associated with fronts because there are no fronts in the Tropics, their third feature. All air is warm there, and the boundaries between adjacent air masses, if detectable at all, differ only slightly in temperature and water vapor present. Hurricanes are usually much more intense than midlatitude storms, their fourth feature, especially in their wind velocity and rainfall. And finally, their frequency has no regularity; many may form one year and nearly none the next.

Somewhere in their journeys, as they wander farther from the equator, they take part in the prevailing westerly winds that form the trademark of the temperate regions in both hemispheres. They turn sharply or gradually toward higher latitudes and take up an easterly motion. When they weaken over land or cool water, they resemble temperate storms; they lose their circular form and assume irregular shapes, they may merge with a front or another storm, and they decay by losing wind and, later, precipitation.

As with temperate storms and low pressure areas, their circulation is always counterclockwise north of the equator and clockwise south of the equator—one of the few invariables in meteorology.

Storms forming in the temperate latitudes come in all sizes and shapes—from little rain showers that pass by in a few minutes to slow massive things lasting several days. Large, slow storms usually form along a warm or cold front, whereas the smaller storms often emerge in the middle of a homogeneous air mass if conditions are right.

In the summertime, highs and lows are large and slow; they drift by, often taking days, even weeks, to do so. In midsummer, fronts are weak because the weather can range only between the hot and the very hot, with little difference between them. Best known of the highs in eastern North America is the deadly and infamous Bermuda High, so named because it is marked by hot and usually humid air centered in the Atlantic Ocean near Bermuda. The Bermuda High is the largest and most powerful of all kinds of weather systems that confront the United States, and it is also the most lethal, killing many more people than do tornadoes and hurricanes, for example. When heat and water vapor build to a certain point, thunderstorms break out; the phrase "scattered showers and thundershowers" is part of almost every forecast in July and August.

This system can assault us in April and May; but at that time of year, cold air is still present and can invade, bringing the brisk freshet weather of spring. Fronts between the tropical humid maritime air and the cooler air masses of Canada can divide various types of air as the disparate air masses give rise to strong fronts in between. An incoming cold front in May can drop the temperature and the dew point by as much as 20 to 30 degrees; in July, a difference of 5 to 10 degrees is more common. May is the month of most frequent violent weather in the forms of thunderstorms and tornadoes. Across the Midwestern plains comes the most violent air experienced anywhere in the world: Tornadoes can spawn winds of 200 to 300 miles per hour, and hailstones the size of baseballs can wipe out a herd of cattle. Few structures can withstand such force, but fortunately these twisters are of limited areal extent, often only a few city blocks; they usually last about half an hour. Several tornadoes are often present in a single storm.

Everyone has experienced a tornado, at least since 1939, when L. Frank Baum's *The Wizard of Oz* was filmed. Although called a cyclone in that movie, the storm that blew Dorothy and Toto from Kansas to Oz was a realistic simulation of a tornado. Tornadoes are the most violent parts of thunderstorms, often stretched along a cold front in a squall line. Fortunately, tornadoes form relatively infrequently.

Rarely in the summer, but frequently in the colder seasons, large storms move in a generally easterly direction; these are more properly labeled cyclones, an alternative term for a low. Wintertime air masses are much smaller and move much faster than summer air masses; sometimes, several lows and highs pass a particular spot within a week. Whenever precipitation can fall as snow, storms generate at the fastest rate of all. A warm front, a wide, lazy system, can be marked by gentle rain or snow lasting most of the day; the cold fronts are sharper, faster-moving systems with fierce wind, rain, or snow lasting only a few hours. Unlike tropical storms, temperate storms are irregularly shaped, frequently with patchy regions of precipitation.

The role of global warming in the formation and intensity of tropical storms has been examined with care. These storms intensify above all over warm waters—the warmer, the better. With some seas showing evidence of a rise in their surface temperatures, hurricanes can only become more frequent and of greater intensity. Furthermore, a warming trend can extend the critical temperature of 27°C (80°F) to more northerly waters, allowing these storms to persist in intensity farther north than is now commonly possible.

In November 1998, hurricane Mitch appeared to be a harbinger of things to come. With maximum steady winds reaching 180 miles (290 kilometers) per hour, at times with gusts to over 200 miles (322 kilometers) per hour, Mitch became one of the four fiercest Atlantic storms on record. It devastated much of Central America and parts of the Yucatán Peninsula, causing at least 10,000 deaths, the highest number of fatalities ever caused by any single storm in the history of the Western Hemisphere. Mitch gained the curious property of intense winds soon after its formation, but as it passed over Central America, it slowed and weakened to a rather average hurricane. Then Mitch stopped right over Honduras and Nicaragua, dumping unbelievable amounts of rain before moving on, with some regions of Honduras getting over 1 meter (40 inches), of precipitation. The infrastructure of both countries was all but destroyed. One superstorm cannot set a trend by itself, but it may well be a portent of weather to come.

The 1999 hurricane season was predicted to foment hurricanes of great frequency and severity. The hurricanes Dennis and Floyd were not as costly as they could have been, although they ended a severe drought across the northeast.

Extratropical midlatitude rains are likely to be enervated by a warming trend. The spring rains and frequent summer showers tend to dry up and pass by without releasing their rainfall, especially during the growing season. Increased drought over the great grain-growing regions of North America seems the most probable outcome. These growing regions play a critical role in the food supply for the world's rapidly growing population; any shortfall would create human misery and economic chaos throughout the world. Wind-battered coastal regions and repeated dust-bowl conditions in the interior threaten to blight our future as temperatures continue to rise. Blizzards continue unabated; global warming has not affected their frequency or their severity in many regions of North America and Europe—at least not yet.

8

WHEN TIME
STOOD STILL:
THE ANCIENT HISTORY
OF THE CLIMATE

We live, by common agreement, in a time called the Holocene Epoch, which began about 10,000 years ago; it is the time of the present inter-glacial period, the time since the glaciers retreated and the last ice age ended. It is also the time when civilization began, although Homo sapiens, with his large brain, came onto the scene perhaps 100,000 to 200,000 years earlier. Before the Holocene came the Pleistocene Epoch, lasting about 2 million years, and during which one ice age followed another with some regularity. The weather history of the Pleistocene—at least the last 500,000 years of it—is becoming well known. Its secrets have been discovered by means of ice cores burrowed into the ice caps of Greenland and Antarc-tica. This chapter describes the climate of those years, and the next one covers our Holocene period.

During the Pleistocene Epoch, the processes for the extraction of carbon dioxide from our atmosphere were still going on; just when the present low level of carbon dioxide in the air was reached is not yet clear, but it had to have been in earlier remote geological times. Evidently, since that point, natural sources continuously or sporadically replenish the carbon dioxide removed by the organic and inorganic processes mentioned earlier.

Many such sources of replenishment are known, some bordering on the catastrophic; for example, at Gerolstein, in the Eifel mountains in Germany, natural fountains release water mixed with carbon dioxide at high pressure.

Another better-known case is that of Lake Nyos, lying in an extinct volcanic crater in the West African nation of Cameroon. In 1986, the lake suddenly released an enormous amount of carbon dioxide into the air, suffocating thousands of people and animals. Because carbon dioxide is heavier than air, it flowed down the sides of the mountain into the valley, pushing the air aside. Today, Lake Nyos and several other lakes in extinct craters in the same area are regularly monitored to prevent further tragedy. Scientists have recently found that the release of carbon dioxide from the region surrounding Mount Aetna, Sicily's perpetually active volcano, is currently replenishing about 10 percent of the annual worldwide loss of this gas.

That carbon dioxide, scarce as it is, can be such an effective absorber of infrared (heat) radiation means that vegetation can act as a temperature regulator. If carbon dioxide increases, the resulting greenhouse effect will raise temperatures and lead to faster and more efficient evaporation of water: More water will be cycled through the air and fall to the surface as precipitation. Increased precipitation and more abundant carbon dioxide favors vegetation, which responds by extracting carbon from the air at a faster rate. Conversely, a shortage of carbon dioxide leads to lower temperatures, less precipitation, and a lower growth rate for vegetation, thus helping to conserve carbon. In both cases, the status quo is restored. These are long-term fixes, though, and cannot be looked upon as a quick relief from global warming.

Carbon is a major constituent of all life forms; thus, it is bound up in organic matter in the oceans. Only recently have scientists realized that because microscopic plankton can temporarily store an excess of atmospheric carbon dioxide, they play a significant role in the carbon story. Plankton are also sensitive to temperature and to the varying ultraviolet radiation from the Sun, changing as it does with the variation of stratospheric ozone.

Might plankton be used, through enhancing its growth, to extract carbon dioxide from the air? The growth of phytoplankton (plankton with plantlike characteristics) is not only affected by the water temperature but also by the availability of iron and other nutrients. As part of an experiment in the Pacific Ocean, scientists seeded an area of several square kilometers with dissolved iron; a spectacular blooming of phytoplankton resulted, accompanied by a significant loss of atmospheric carbon dioxide in the area. Because the consequences of a large-scale application of this method are unclear, controlling the present manmade increase of atmospheric carbon dioxide by this method is too risky.

The carbon dioxide stored in the oceans is in part regulated by the abundance of phytoplankton, which, in turn, is regulated by the availability of iron; thus the variability of iron input into the oceans may have contributed to the variation in the abundance of atmospheric carbon dioxide in the past. Iron is transported into the oceans by rivers that flow down iron-bearing mountains and into the sea. An additional source of iron may come from the bombardment of the Earth by extraterrestrial iron meteorites; such an influx may have been variable during past geologic times.

These stabilizing mechanisms can work only up to a critical point; when that point is reached, the land surface and the oceans become saturated with organic life and no more carbon can be stored by vegetation or plankton. If the abundance of carbon dioxide in the air increases further, the greenhouse effect takes over and temperatures rise. The release of gas by volcanic activity, a large source of carbon dioxide, has not been constant during the time organic life has been present on the Earth. The tectonic plates are continuously in motion, producing continental drift and its attendant volcanic activity along the cracks between the plates. New volcanoes form constantly and others become extinct; volcanoes thought to be extinct sometimes come back to life. The release of carbon dioxide must have undergone considerable periodic and erratic variations over the ages.

Not all potential climate-altering mechanisms are terrestrial. The total radiation flux of the Sun has been virtually constant during the entire geological history of the Earth. The energy-producing process by which the Sun shines results from the fusion of hydrogen into helium in its core. The Sun is a self-stabilizing mechanism, a furnace made from its own fuel; it may vary with periodic oscillations, as some other stars are known to do. But the Sun varies in radiation only with a small amplitude, if at all.

The sunspot cycle, with varied activity over about eleven years, is the only confirmed variation of the Sun. The sunspot cycle has been examined as a possible cause for the swings in the Dow-Jones Index, for baseball statistics, and for most everything else, all without success. None of the many studies have convincingly shown that sunspots influence variations in climate, rainfall, or crop growth.

Climatic variations of any significance caused by solar irregularity could only hold for time scales lasting over hundreds or thousands of years or more. Because organic life has been continuously present for several billion years, any variations of the solar energy flux must have had a relatively small amplitude, not large enough to contradict the geological record. No method is yet known to confirm the existence or absence of solar variations over long geological time scales.

The ice ages remain the most striking climatic variations during the more recent history of the Earth. Most of the traces left by the masses of ice that cover the continents at irregular intervals are vulnerable to erosion. Because these traces are wiped out in a short geological time, we do not expect to find many residual signs of glacial activity—moraines, glacial lakes, valleys shaped by glaciers—that were formed earlier than the Pleistocene Epoch. Traces more than one billion years old have been found; however, large ice masses in the Tropics appear in just the areas we would least expect them. As layers of ice more than a kilometer thick slowly glide over a rocky surface, they mark it with characteristic scratches that, if covered by new deposits, can survive almost forever. Such scratches have been found on the oldest rocks that were once exposed at the surface.

The motions of the continents over the last several hundred million years are well known. In South Africa, one find of rocks covered with glacier scratches told us where on the globe South Africa was once located. Furthermore, magnetic lines frozen into the rock have fixed the past locations of the Earth's poles and the geographic latitude of this South African site. The surprising result is that the site was at a latitude of only about 10 degrees north of the equator, right in the middle of the Tropics. During that particular ancient ice age, the Earth was covered in ice from both poles almost to the equator. Earth must have looked more like a giant snowball than the planet we know. Little else is known about the climate of the time, but this discovery shows us that catastrophic climate changes did happen here from natural causes; and we can only speculate about what those causes were.

Most of the conclusions about historical climates are derived from the study of fossils found in sedimentary rocks all over the world. By the nature of their formation, sediments preserve a chronological order, their time scales now well established. In arctic regions, fossils of fauna and flora are different from those in the Tropics. A comparison of fossils with present forms of life allows us to draw conclusions about the climate at the time and site to which the fossils belong.

Recent techniques, based on the relative abundances of isotopes of certain chemical elements, have provided detailed and precise information about some of the climates of the past. Isotopes of an element are atoms that are chemically indistinguishable from each other but have different atomic weights. In some chemical reactions as well as in some physical processes (such as the formation of ice crystals), normally heavier isotopes may be favored or disfavored with respect to the most common isotope of an element, the difference being a function of the temperature. The relative isotope abundance in the products of chemical reactions or in ice crystals may tell us more about the ambient temperature prevailing when the reaction took place or when the crystal was formed. The isotope method requires specific circumstances, which are not always fulfilled. We cannot yet use this method to obtain a continuous temperature record over all geological periods, but the method is an important complement to the results derived from fossils of the last one million years or less.

For 550 million years, vertebrates have left a record of fossils that reveals a continuous evolution and diversification of life forms. The fossil record still has many gaps, and many mysteries remain to be solved; we can safely conclude, however, that a spontaneous appearance of advanced life forms on the Earth has never taken place, despite the continuing attempts of creationists to prove the opposite. By contrast, striking examples of the simultaneous disappearance of large numbers of species are recognized; the best known among these "mass extinctions" brought about the disappearance of the dinosaurs, together with all other large land animals, about 65 million years ago. From 50 to 70 percent of other species then extant also vanished.

This event, known as the *K/T boundary,* is named for the point that separates the Cretaceous (sometimes spelled in German with a K, therefore the designation K/T) and Tertiary periods; it is identified by a marked division between the layers of sediments deposited in the Cretaceous period from later Tertiary deposits. Four earlier mass extinctions are recognized; one of these separates the Permian and Triassic periods and is thus referred to as the P/T boundary. This event, believed to have happened some 230 million years ago, may have wiped out 96 percent of all marine species, and evidence exists for at least three other similar major events, which occurred about 440, 365, and 210 million years ago. Many anthropologists postulate the existence of a sixth mass extinction, caused by our own species. It began about 10,000–12,000 years ago, when the first human settlers of North and South America hunted to extinction the many species of large mammals (mammoth, mastodon, ground sloth) of the New World. Extinctions continue as we spread across the globe and top 6 billion in population; we endanger and deplete the grazing land necessary for the survival of the remaining largest land creatures such as the elephant, the rhinoceros, the hippopotamus, and the giraffe.

Drastic climate changes are among the most certain results of all five accepted mass extinctions, each of which was probably followed by a burst of evolutionary development of new species to fill the vacancies left by those that vanished. Isotope analysis supports this hypothesis. During the last twenty years, the extinctions have been extensively investigated, bringing to light the spectacular discovery that the element iridium shows an excessive abundance at the K/T boundary, being present in amounts as much as one hundred times its normal abundance in many sites scattered around the world. To this we may add the more recent discovery of microdiamonds in the same layer. Iridium and microdiamonds, rare substances on the surface of the Earth, are abundant in meteorites of cometary origin. The K/T event is the most studied of the five major disruptions because it occurred so much more recently than the others and less of its record has vanished through natural processes.

These findings, particularly the overabundance of iridium, led the Nobel laureate physicist Luis Alvarez, his paleontologist son Walter, and two other scientists to formulate the now-famous *impact theory* (made public in about 1980). According to the theory, the impact onto the surface of the Earth of a comet or asteroid, perhaps 10 kilometers (6 miles) across, ejected a huge cloud of dust into the atmosphere all the way into the stratosphere and beyond; the result was an atmosphere almost opaque to incoming solar radiation. A number of theoreticians have examined possible effects of such a high dust contamination of the air, although their hypothetical source of dust was not a natural catastrophe but a full-scale nuclear war. They concluded that a "nuclear winter"—the entire planet plunged into intensely cold conditions for several months—is to be expected after an exchange of even a minor portion of present nuclear arsenals. These conclusions are even more valid for the consequences of an impact of a major object 10 kilometers (6 miles) in diameter, which would be thousands of times more destructive.

The Alvarez theory soon became one of the most widely known and popular theories of the century in the natural sciences; it formed the basis of at least two blockbuster movies in 1998 alone. The popularity came about mostly because the theory holds that a sudden catastrophic occurrence killed all the dinosaurs at one time, not a slow succession of alterations in the geologic or climatic conditions of the day (although this may have been a contributing factor).

This cataclysm helped resolve the long-standing persistent debate between *catastrophism* and *uniformitarianism*. These are terms for the doctrines that hold that nature can change relatively suddenly and massively; alternatively, geologic and other changes come about only very slowly through gradual evolutionary processes. Since the early nineteenth century, when the great geologist, Charles Lyell, proposed it, the uniformitarian concept had prevailed; in recent decades, evidence for short and calamitous events has grown along with the popularity of catastrophism. The impact theory helped revive catastrophism as an alternate mechanism for change. Its impact brought to human consciousness the possibility that we, too, could be wiped out in a flash, and also that we have (or nearly have) the means to prevent it.

Many scientists disdained the theory at first, partly because of their residual belief in a measured pace as the only medium of change and partly because no site could be identified with the impact. But since the introduction of catastrophism, acceptance for it has grown. In 1991, the long-sought site of the impact was discovered. This smoking gun, in the form of a huge vestigial crater, is known by the Mayan name of Chicxulub. The crater lies at the edge of the Yucatán peninsula of Mexico and is partly flooded by the Gulf of Mexico. Chicxulub appears to have a diameter of more than 200 kilometers (120 miles), which corresponds to a diameter of about 10 kilometers (6 miles) for the impacting body. Ejecta with microdiamonds, spherules, and abundant iridium have been found around it. Its age coincides exactly with that of the K/T boundary, 65 million years ago.

Today, it is widely accepted that a sizable asteroid of perhaps 5 to 10 miles (about 8 to 16 kilometers) in diameter and orbiting on a northerly course piled into the coast of the Yucatán Peninsula and the Gulf of Mexico at a relative speed of about 20 miles (32 kilometers) per second. The damage from this greatest cataclysm of the past 200 million years or more is hard to imagine. Lurid descriptions of the sudden and tragic end of the entire Mesozoic world appear in recent popular books by Walter Alvarez, Richard Leakey, and Peter Ward. The Mesozoic Era, framed by the P/T and K/T events, is likened by Alvarez in his *T.Rex and the Crater of Doom* to the Middle Earth of J.R.R. Tolkien's *The Lord of the Rings*.

We can be grateful that we have no experience for imagining a tragedy that blew that alternate world away forever. No legends like those of the Biblical flood and Atlantis grace our collective mythos of an event of this magnitude. Even an all-out nuclear war pales next to the Chicxulub event. It is the incredible velocity of this Mount Everest-sized thing that gave it such a force. Rock behaves in unfamiliar ways when struck at such a speed. As shock waves compress it, the rock can vaporize or just fly apart on the decompressing rebound. The object vaporized in about a second while blowing a hole 40 kilometers (25 miles) deep, which then broadened into the crater we know. Many forest fires were ignited (as would happen in a nuclear war for obvious reasons). Smoke and ash could well account for an opaqueness of the air; indeed, traces of soot were detected right at the deposits laid down at the time of the K/T boundary.

Alvarez states: "As the rapidly vaporizing comet wreckage was carried forward into the growing crater, the shock wave curved back up to the surface and spewed out ejecta—melted blobs and solid fragments of target rock—upward and outward on high, arching trajectories that flung them through the thin outer fringes of the atmosphere and beyond." The fireball created by the blast, incomparably greater than any nuclear device, blew right through the atmosphere. Shocked layers of limestone released tons of stored carbon dioxide along with huge pieces of rock. The crater, temporarily some 40 kilometers (25 miles) deep, was too large for the remaining crust to support it; its center bounded upward in a new wave of chaos. Because the death missile came in from the south, North America took the full force of the blast. Nothing survived in the region now known as Texas—nothing.

In that apocalyptic carnage, tsunamis and tidal waves, perhaps a kilometer high, charged around the world. Cubic miles of dirt and dust were thrown into the stratosphere, far more than would rise from a total nuclear exchange emptying all of our atomic arsenals. There the dust remained for months. The entire globe was plunged into a cold winter; all but 1 percent of the available sunshine was blocked by a cloud-enshrouded atmosphere that must have resembled the present atmosphere of Venus. Such an event is sometimes referred to as a mass extinction because so many species pass into oblivion almost at once.

During the several thousand years that passed before normal conditions returned, more than half of all species of plants and animals vanished, including all the remaining dinosaurs. These great saurians had ruled the land for 140 million years; their disappearance may have allowed tiny mouse-sized mammalia to develop and eventually dominate the remade world.

Strong support for the impact hypothesis came though the detection of *shocked quartz* in the boundary layers. This type of quartz forms under high temperatures and pressures. Such conditions can occur only through an impact at high velocity, as would be expected in a collision between the Earth and a comet or an asteroid. But what we have described so far may not prove the impact hypothesis beyond a reasonable doubt. Increased volcanic activity could cause fairly similar effects—including the enrichment of iridium because this element is more abundant in the interior of the Earth than at its surface. The formation of diamonds is also related to volcanic activity.

Other evidence favors an alternate explanation for some of the mass extinctions. Most of the dust stirred up by an impact would not have remained in the air for long; after a few months, most of it would have settled onto the ground either by its own weight or through the mechanism of condensation nuclei. Only microscopic particles would have lingered in the atmosphere. Smoke and ash, too, would not have sustained a drastically different climate for a prolonged period. It is difficult to imagine that so many species would be completely wiped out within a few months of an adverse climate; this is particularly true for vegetation, also drastically affected during the mass extinctions. The alternate scenario, then, suggests not a single impact but rather a series of impacts spread over a considerable period; such events could maintain a highly contaminated atmosphere for many years or even centuries. Possibly both hypotheses, that of the single impact and that of ongoing volcanic activity, could be combined: A major impact could have triggered violent volcanic eruptions that continued over a long interval.

Could such a disaster happen again? Yes, of course it could—we still have large chunks of stuff circling round the Sun. But we are such a widely spread animal that it is unlikely we would all die. Every trace of civilization would go in a moment, however, and the survivors, if any, would again live a hunter-gatherer existence—if there was anything left to hunt.

The collision of the pieces of Comet Shoemaker-Levy 9 with Jupiter in July 1994 was the greatest catastrophe humans have ever witnessed in the solar system. It alone compares in violence with the K/T disaster, and it brought us up on such an event's happening on Earth today. This and the growing acceptance and attractiveness of the Alvarez theory have led to a remarkable new interest in asteroids and comets—the often ignored debris of the solar system left over from its creation. The hit movie *Jurassic Park* and other media events have embellished at least one supposed consequence of the great K/T event and the fascinating theory behind it. Had the dinosaurs lived, the speculation goes, might some of them have evolved into beings with an intelligence on a par with our own, and if so, might they have prevented the rise of Homo sapiens? Science fiction has long dwelt on such themes, as well as others dealing with reptilian intelligence that might live somewhere in the galaxy.

•   •   •

Although almost convincing evidence favors the impact theory, a number of empirical facts cannot be readily explained by it. Certain types of amino acids are present in meteorites, most of which form the remains of disintegrated comets. The acids have been detected at the K/T boundary; however, they are also present below and above the boundary, thus indicating that the Earth had been exposed to deposits of cometary debris for a long time. Temperature studies using isotopes also show that the cooling may have begun 200,000 years before the K/T event and produced a maximum warming 3.6 million years afterwards. Another difficult observation to reconcile comes from the so-called Deccan traps, a large basalt formation in the Himalayas. Basalt forms from hot, liquid material upwelling from the deep layers of the Earth. The Deccan traps show there have been a number of these types of floods. That they appear to be nearly periodic is striking; they also seem to coincide with some of the major extinctions.

Two well-known conditions apply to the argument. One is the apparent periodicity of events, indicating cyclic behavior that might be caused, in one scenario, by a companion star's orbiting the Sun at varying distances and so perturbing comets from afar toward the Sun and Earth every 30 million years or so. But such a scenario does not stand up to closer examination. The second condition is well known to researchers in many fields. A number of events that happened at the same instant in the past are much more likely to show different times of occurrence than are events with a spread in time in their formation to appear by chance to have happened simultaneously. Any dating procedure carries an error in the process; although such errors give rise to a spurious scatter or spread in time, the opposite never occurs. How often would events that took place between, say 60 and 70 million years ago all appear to line up by chance to 65 million years ago?

Not all explanations of the other earlier mass extinctions are astronomical. A purely geophysical mechanism has also been suggested for their cause. This mechanism relates to the stability of the so-called D-layers deep inside the Earth's interior. Heat developed by radioactive processes in atomic nuclei may cause the lower strata of the D-layers to become lighter than the overlying strata. The lighter material will then periodically puncture the layer above it and break through, upsetting everything above it all the way up to the solid crust. Such an event can readily cause motions of the tectonic plates and trigger volcanic activity. Once the excess pressure in the lower D-layers is relieved, normal stratification is restored; the volcanic activity may cease some time thereafter. Such a process may well be semi-periodic. One argument against this hypothesis is that it cannot account for the excessive iridium abundance found at the K/T boundary. This, however, may not be the case because iridium is more abundant at great depths and micro-diamonds and shocked quartz could have their origins there as well.

Fortunately for our species, not many asteroids of the size of Chicxulub are still around, and most of those that are have well-determined orbits. And few icy comets are this large, although the spectacular Comet Hale-Bopp, which appeared in the skies in 1997, is indeed of the required mass to finish off the lot of us. Smaller celestial detritus, perhaps 1 kilometer or 1 mile across and possibly able to take out a continent at one shot, are yet plentiful. Some day, whether with nuclear missiles at the ready or with other less violent solutions now on the drawing boards, we may manage to defend ourselves. A nuclear detonation alongside an errant comet or asteroid could blow it into slightly different orbit, thus causing it to miss the Earth. Other tricks to accomplish the same result less destructively may also turn out to be as practical. For any deterrent to work well, the warning time, the time available for it to be brought into position alongside the errant object, is all important.

In March 1998, the media seized upon a report that an asteroid of about 1 kilometer would pass only 50,000 kilometers (31,000 miles, or one-eighth the distance to the Moon) from the Earth in October in the year 2028. The prediction caused a feeding frenzy in the media, but it encouraged astronomers to look for past observations recording the asteroid's position; thus the epoch difference between first and last observations was lengthened, which lead to a major improvement in the determination of the exact orbit. Fortunately for us, observations were found in archival material that showed the most likely distance at that future time to be almost 1 million kilometers (620,000 miles), more than twice the distance of the Moon. In coming years, scientists will further improve the precision of the prediction.

As a result of this perceived threat of a collision, the search for all significant objects in orbit around the Sun has undergone a resurrection. The largest problem in the scenario of a deterrent, nuclear or otherwise, lies in the shortness of the period over which we know that another K/T event, or even a Sodom-Gomorrah-level of disaster, is headed our way. A complete census of at least the largest asteroids will help prepare us. When all is said and done, however, the insurance of extreme infrequency is almost all that is left to us.

A strike by a comet or asteroid is a unique event in two ways. First, it is about the only kind of disaster that could wipe out all civilization, or even all life. No earthquake, volcanic eruption, hurricane, or tsunami has that capacity (although some have speculated that the P/T event was caused by massive volcanic activity in Siberia). These disasters are today local or regional. The destruction from a large impact would be worldwide, a global cataclysm.

The second singular feature of such an event lies in our potential ability to deflect the intruder and prevent the catastrophe. No way can humankind prevent an earthquake or divert a volcanic eruption of lava, or even slow down the winds of a tropical storm. But we can prevent a collision with an asteroid or comet; we lack only the funding to do so.

Two main kinds of junk exist in space with the potential once again to change the course of life and its evolvement through mass extinctions: the asteroids and the comets. Fundamental differences distinguish the two groups. Asteroids, or minor planets, are the small, mostly rocky bodies that circle the Sun directly. Most are found about midway between the orbits of Mars and Jupiter; collectively they are known as the *asteroid belt.* The largest, Ceres, is about 600 miles in diameter (1,000 kilometers), about twice the size of the next two largest. About thirty larger than 60 miles (100 kilometers) in diameter exist, and three hundred are larger than 6 miles (10 kilometers), the size of the despoiler of the dinosaur's world. Most orbit where they belong, in the asteroid belt, and it is unlikely that any would be gravitationally perturbed into an orbit in conflict with ours. Continued observation assures us that we will know well in advance should any problem of this magnitude arise. When we consider smaller asteroids, the problem worsens. About 6,000 asteroids are known and named, their sizes being mostly from 1 kilometer to 1 mile across or a bit larger. But as many as ten times this number may exist somewhere out there. L. Niven and J. Pournelle graphically portrayed the destruction from an object of about this size and its impact on civilization in their novel *Lucifer's Hammer.*

Comets, the other large constituent, are part of the outer icy solar system. Along with Pluto and most of the satellites of the major planets, they are mostly composed of water ice. *Dirty snowballs* is the well-known description of them because tiny rocky and dusty particles are often embedded in the ice. As a comet moves into the inner solar system (roughly the asteroid belt and closer), the increasing influence of solar radiation heats it and melts and evaporates some of the ice to the point where the tiny dust particles are freed and then repelled away from the Sun. The freed dust particles form a great tail, straight or curved in appearance, that forms the spectacular part of a comet's show. Comet tails always point away from the Sun, regardless of the direction of the comet's own orbital motion.

A second type of tail, usually fainter and more bluish, is formed of ionized gaseous material that points more directly away from the Sun. Comet Hale-Bopp at its brightest exhibited both kinds of tails, but the blue ion tail was visible only to those fortunate enough to observe the comet well away from light pollution.

The comet factory lies beyond Pluto, out to as much as 10 to 50,000 times the Earth's distance from the Sun. The factory is called the *Oort Cloud* after its discoverer, and consists of thousands of small dirty snowballs, and a few larger ones. They hang around out there unless and until some external force, a collision or gravitational perturbation, throws them into orbits directed inward toward the Sun and planets. That is when the trouble can begin.

The Earth remains relatively unscarred compared to the Moon. Our world, being the larger, was much more bombarded than the Moon; but because Earth is geologically and seismically active, it has the means to resurface itself and cover the countless craters of the past. Ocean covers two-thirds of the surface, and plate tectonics create a continuous process of continental drift, with overlapping and subduction ever occurring at the borders between the plates. Venus, and to a much lesser extent, Mars, are seismically active or have other means for obliterating old craters; and the four giant planets are shrouded in thick atmospheres. But except for these and, of course, the Sun, all other objects in the solar system remain littered with craters from that early bombardment. Even the two tiny Martian moons, far and away too small for geological and volcanic activity at any time in their history, are cratered to the full, thus confirming the meteoritic origin of most craters everywhere.

Thus only very recent strikes remain to be seen on the Earth. Two events mark the greatest of what recent history has to offer. About 25,000 to 50,000 years ago, northern Arizona was struck by a meteorite or small asteroid of high iron and nickel content. It was probably about 150 feet (50 meters) across, and weighed as much as 1 million tons. It moved 10 to 15 miles per second just before it struck the Earth, hurtling from the stratosphere to the ground in its last second of life. Most of it shattered or vaporized upon impact, but not before it gouged out a crater in the crust, about

4,000 feet in diameter and 600 feet deep at the center. Many pieces of the object have been found, some weighing as much as a ton, but no larger pieces exist. It is the high rate of speed that gives a projectile like this its great potential for disaster. The crater, known as the Barringer Crater, located not far from Winslow, Arizona, is a major tourist site; it is probably the best-known and most visited crater of its kind anywhere on the globe.

In recorded history, our most devastating celestial visitor was probably a small comet that smashed into remote northern Siberia on June 30, 1908. No human casualties are known from this event, called the *Tunguska Event,* although trees were felled for tens of miles in all directions. The comet hit in the daytime, yet reindeer herders saw the glare and heard the noise of the explosion for hundreds of miles in all directions.

Our most spectacular and widely witnessed event involved no collision at all. On August 10, 1972, a huge glowing fireball, made by a meteoroid as big as a boxcar and weighing tens of tons, passed over the U.S. and Canadian west in the daytime. Many vacationers noticed it as it moved over Utah and up through Alberta, passing over Grand Teton, Yellowstone, and Glacier National Parks along the way. After skimming along in our upper atmosphere nearly parallel to the ground, it proceeded back out into space, not much the worse for wear. Had it landed, it would have made a large crater suitable enough for a tourist site. Somewhere out there this chunk orbits the Sun, now somewhat smaller and on a different path from the one it had before its encounter with the earth.

Smaller meteor falls happen every year. They typically leave fragments weighing a few pounds and do little damage. By definition, a chunk that has reached the ground is a meteorite, but the object in orbit before the fall is known as a *meteoroid.* The boundary between large meteoroids and small asteroids is not defined. Meteors are those "shooting stars" that are too small to reach the ground; they burn up in the atmosphere as friction heats them to thousands of degrees. Most are no larger than a grain of sand, even before their fiery encounters with our atmosphere.

Collisions with objects of the size of the one that caused the K/T event are rare, perhaps one per 100 million years being the average rate. Smaller ones a mile in diameter may come in with a frequency of one in 1 million years. The odds are still heavily on our side.

Comets can be divided into two groups on the basis of their orbital periods. Some have periods of years or decades or even several centuries, and others appear to have periods lasting thousands of years or longer and have appeared no more than once in recorded history. Although all are thought to orbit the Sun, only those with the shorter periods are referred to as periodic. All comets have orbits with a large eccentricity, which means that at one end of their orbit they come close to the Sun and the Earth, but at the other end the periodic comets pass well beyond the orbit of Jupiter and the others travel to far greater distances. For example, Halley's Comet, much the best known, has a period of seventy-six years and an orbit that extends just beyond Neptune's orbit. Each time a comet passes close to the Sun, it heats up and evaporates part of its surface, which is then blown away by the solar wind, thus making a tail. The scattered material does not fall back onto the main body of the comet but disperses into space and thus is lost; this means that during each passage close to the Sun, a comet loses material and eventually disintegrates completely. Several well-observed comets have actually disappeared during historical times for this reason. From estimates of the size of comets (which have become well known since the encounter of the spacecraft Giotto with Halley's Comet in 1986), one can estimate through how many passages a comet can last. The maximum number of passages turns out to be a few thousand at most. This means that all the presently known periodic comets, including Halley's, will come to an end in well under 1 million years.

The solar system is already some 4.6 billion years old, and it is extremely unlikely that just now we are observing a transitory phenomenon. It is more reasonable to assume that comets have been around during the entire lifetime of the solar system. As we described earlier, the present view supports the existence of a large reservoir of comets in the Oort cloud. The proto-comets in the Oort cloud are supposed to move in near circular orbits and thus remain in their distant domain, far beyond the orbits of the planets, but still a part of the solar system.

By a still unknown mechanism, some of these proto-comets are perturbed into orbits that cause them to fall into the interior of the solar system. If they happen to come close to one of the major planets, most likely Jupiter because it is the largest, their orbits can be altered in such a manner that they remain permanently in the vicinity of the system of the planets. In other words, they become periodic comets. Some will collide with one or another of the planets and thus vanish, and the final orbits of the rest will be such that they never come close enough to a planet for a collision or encounter to take place. In this case, they may last for thousands of years until they disintegrate or fall into the Sun.

The great dinosaurs, including even mighty Tyrannosaurus Rex, ended in a reign of fire. Homo sapiens came close, and may again come close, to repeating the event through his own nuclear irresponsibility. Now we can avoid this calamity from our own actions, and soon we will be able divert nature's blunders as well. It is not too early to take preventive measures against catastrophe from either direction.

9

HERE COME THE GLACIERS

The ice was here, the ice was there,
The ice was all around
It cracked and growled and roared and howled,
Like noises in a swound!

SAMUEL TAYLOR COLERIDGE,
THE RIME OF THE ANCIENT MARINER

ANY RECONSTRUCTION OF HISTORY is likely to be most precise in the near past and become less certain as we go back in time. This is certainly the case in climatology. Although we know yesterday's weather well, and even the last century's fairly well, we know less about the weather in past centuries, and only now are we confident about climate during the ice ages.

More and better methods exist today for the reconstruction of the climate in the Pleistocene epoch than for the climatic history of earlier times. The Pleistocene epoch is defined as the last 2 million years or so; its characteristic spectacular ice ages have left their marks across the world. Because each ice age erases most traces of the previous one, only the last of them is well documented. Nevertheless, evidence exists for several others and they, too, have been dated precisely.

When viewed in the context of geological time scales, ice ages are comparatively short-lived; each of the most recent ones lasted for about 100,000 years. This may not seem long, but the most recent one or two ice ages encompass all of the existence of Homo sapiens sapiens and its close cousin, Homo sapiens neanderthalensis; these are the two big-brained apes, each having about 1,400 cubic centimeters of gray matter, far more than that of earlier and less intelligent hominids such as the Australopithecus species. Anthropologists believe that the great increase in brain size and capacity came about during two relatively brief periods, with little development between them. The first increase took place about 1 million years ago, and the second, which more than doubled brain size, occurred some 100,000 to 200,000 years ago, or one to two ice ages in the past. Those who continue to push the oxymoron "creation science" dispute these conclusions, but offer no evidence for their beliefs.

During these cold periods, called *glacial periods,* the polar ice shields and pack ice in the oceans extended much closer to the equator than they do today or than they did during the briefer, warmer interglacial periods of the past. At the end of the last glacial period, about 10,000 to 12,000 years ago, the north polar ice extended over most of Canada and well into the United States. It covered northern Europe and reached into central Europe; it also covered most of Siberia. At the same time, glaciers originating in mountainous areas, even those in the Tropics, were considerably larger than they were during warm periods (such as today), and some formerly glacial areas have disappeared altogether. Much of this ice must have been several hundred meters or even several kilometers thick, comparable to the present ice shields on Greenland and in Antarctica.

The Pleistocene epoch is defined as ending at the close of the last period of glaciation, the last ice age, when the Holocene or Recent epoch began, about 10,000 years ago, or 8000 B.C. In 11,500 B.C. or near, the glaciers that had dominated much of America, Europe, and Asia for so long began a retreat that was completed well within 3,000 years. With the glaciers' retreat, ocean levels began to rise because less of the world's water was locked up in frozen form. The warming and melting continued, and sometime after 7000 B.C., the English Channel washed over the land bridge from Britain to the coast of France and connected the Atlantic Ocean with the North Sea, as shown in Figure 9.1. The Thames River was no longer a tributary of the larger Rhine River flowing across grasslands into a diminished North Sea, and Britain began its insular period, with a guaranteed maritime climate free of most extremes in temperature. In North America, the Great Lakes were formed as the glaciers retreated.

The most recent ice age, lasting from about 100,000 years ago until the start of the Holocene epoch, is known as the *Wisconsin Ice Age* in the United States and as the *Würm Ice Age* in Europe. A small portion of southwestern Wisconsin remained a kind of hilly island mostly free of glaciers, though surrounded by them on all sides, hence the American version of the name. Remember that the prevailing temperatures and therefore the extent and thickness of the ice sheets were not constant, not at all,

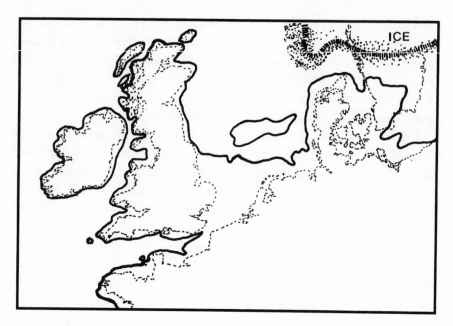

FIGURE 9.1   Map of the British Isles about 8000 B.C. showing the lower coast-line at that time (solid line) with the present coastline (dotted line).

during the 90,000-year Wisconcin glaciation. We know that an interstadial period occurred about 30,000 years ago, when the ice age let up a little. This milder time formed the setting for Jean Auel's popular *Earth's Children* series of novels.

Interstadial periods bring milder times within ice ages, but they are not warm enough to be classed as interglacial periods, as is our present Holocene epoch. The difference between that interstadial period and to-day's interglacial epoch is considerable, as shown in Figure 9.2 for Europe. One important difference is the extent of reforestation of former ice fields. Soon after the Holocene epoch began, forests of oak and other deciduous trees moved in to cover Europe; reforestation by oak serves as one of the delineations between the two levels of comparative warmth.

The coldest period and most severe conditions of the entire Wisconsin glaciation arose about 18,000 years ago, when Chicago and New York lay

FIGURE 9.2   Map of Europe about 30,000 B.C. during an interstadial period showing glaciated regions.

under a mile of ice; it was then that glaciers made their most southerly advance. From that moment until well into historical times, the ice sheets have been in retreat, the melting ice flowing into the oceans and raising their level considerably (see Figure 9.3).

Before the great warming at the start of our present interglacial Holocene epoch, the continents were not as we know them today. Because so much more of the world's oceans were locked up as ice, the ocean level was lower then by almost 100 meters (some 300 feet). Not only was Great Britain attached to Europe by a land bridge where the English Channel is now but Alaska was attached to eastern Siberia by another land bridge that spanned the Bering Straits; this bridge allowed the first human settlers of the New World to migrate on land from Asia into the Western Hemisphere—and over water in small boats where necessary. Some anthropologists believe that the first humans to settle the Americas did so only after they had acquired the technical skills to thrive in the arctic conditions of Siberia and Alaska. Furthermore, the glaciation of upland areas in Alaska and northwestern Canada blocked the settlers' move south until the years just before the early Holocene, when many of the uplands became passable.

FIGURE 9.3   Map of North America, showing the maximum glaciation of the Wisconsin Ice Age, 18,000 B.C.

The Black Sea was a much smaller fresh-water lake in those days, not yet connected by water to the Mediterranean Sea or to any other part of the world's ocean system. Not until well after our recent epoch began was the connection made. The oceans continued to rise for thousands of years after the glaciers had begun their final retreat, even after the discovery of the amazing cultivation properties of wheat and other grains some 8,000 to 10,000 years ago. This discovery brought an alternate and settled lifestyle to our nomadic hunter-gatherer ancestors through the development of agriculture, and through that, the rise of communities and the division of labor, leading in their turn to written historical records.

By 5600 B.C., these developments were well underway. At that time the Mediterranean Sea rose to fill the Bosporus Strait, the narrow 30-kilometer waterway that now flows past Istanbul. In a matter of a few years, water broke through into the much lower Black Sea in probably the most stupendous cascade ever witnessed by humans. In less than a decade, the Black Sea rose by 120 meters (400 feet), to its present level, and may have provided the origin of the legends of Noah's biblical flood and the similar flood in the Babylonian Epic of Gilgamesh.

The transition from the Pleistocene to the Holocene epoch was a time of great change, not only in climate but in the plants and animals that abounded in Europe and the Americas. As we noted above, great forests sprang up at that time, and giant mammals were not then limited (as they are now) to Africa and southern Asia. In those days, mammoths, mastodons, giant ground sloths, and saber tooth tigers roamed North America and Europe, all to be hunted to extinction about 10,000 years ago, probably by our nomadic forbears moving in from Asia. Scientists now believe that, even after the bronze age replaced the stone age, some mammoths survived on Arctic islands north of Siberia until about 2000 B.C., or after the great pyramids and Stonehenge were built. The notion that some of our forebears shared the globe with giant, hairy elephantine creatures may be novel to us, but Abraham, Cheops, Imhotep, and Gilgamesh all did so. That the oldest living organisms are a few trees that may be as much as 4,000 years old is just as thought provoking. When they were young they, too, may have shared the Earth with the last of the mammoths.

What precipitated the onset of the ice ages? Certainly the primary cause was that the Earth, or parts of it, grew colder. But, as any child would then ask, *why* did the Earth get colder? Several mechanisms, which we discussed in Chapter 2, could have triggered the change. Probably a substantial change in the global temperature was involved. At least two mechanisms can give rise to glaciers and cause the polar ice caps to grow. The more obvious possibility is that summer temperatures were lower than they were during the interglacial periods. If that was the case, naturally, less ice melted and more ice accumulated as time went on; the snows of one winter piled up more quickly on top of snow from the previous winter than they had on bare ground. Another possibility is an increase in snowfall. If that was case, the summer melt could not catch up to the additional ice accumulated during the winter. An increase in the global temperature can cause more water to be cycled through the air, resulting in more precipitation; some of this precipitation could well be unloaded onto the arctic regions in frozen form. The existence of ice ages does not lead to a definite conclusion about the global temperature as a whole.

The seemingly contradictory situation becomes obvious in the Tropics. On the one hand, most of the evaporation there arises from the oceans, the lakes, and the lowlands. The higher the temperature, the greater the water evaporation. On the other hand, because high mountains in the Tropics are covered with clouds most of the time they may never see the Sun during the wet seasons. While the low areas see hotter weather, snow and ice cover on the mountaintops may continue to grow.

Temperature data from many parts of the world must be obtained before we can reach a conclusion about the history of the Earth's climate. Information on precipitation, water vapor content, and the composition of the atmosphere must accompany the more obviously needed temperature data. We cannot be content with measures made only on the ground; we must also obtain measures on the oceans' surfaces and at all altitudes aloft, even into the stratosphere. Our atmosphere is a three-dimensional affair and height is as important as latitude and longitude. Scientists are working in that direction but still have far to go.

As with any one of the proverbial blind men each feeling part of an elephant, when we look at the climatic history of a site, the information we gain is of only limited relevance to the global climate history; worldwide data are needed. Various data-collection methods, some elaborate, are available, and although each applies to a specific case, it contributes to the total evidence. Because we cannot describe the technical details of all of them, we will describe the most relevant. In interpreting the data and conclusions within a common framework, results from what are called *General Circulation Models* are helpful.

Several points of evidence link overall temperature with ocean levels. Even within the Holocene, lake and ocean levels correlate with temperature and help define the relatively warm and cool times. We can estimate the relative warmth of climates in classical times from piers and other harbor construction at Ostia and other port cities of the Roman Empire; climates then were warmer than ours and the Romans built their docks at a higher level.

Broader evidence rests in the ancient shorelines and coral reefs of the past. The local topography shows variations in the extent of the polar ice shield and of mountain glaciers; these variations are traceable from the rocky material the ice carried in its slow, continuous motion. Glaciers in mountainous areas create typical glacial profiles in the shapes of the valleys through which they descend as they gouge out rocks and boulders. Loose material pushed along and piled up at the lower extremes turning the piles into so-called *terminal moraines.* Analysis of the abundance of unstable carbon isotopes in the remains of organic material permit accurate dating of the glaciers' maximum extension. The north polar ice shield was also in a permanent but slow motion as it scraped the surface below; the smoothed landscape left behind is easily recognized on aerial photographs. Thus from the size of the area covered by ice as well as the amount of water in frozen form—water missing from the oceans—we can deduce the average thickness of the ice.

Today, massive shields of ice cover almost all of Greenland and Antarctica. These shields are as much as 13,000 feet (4,000 meters) thick in places and grow year after year as more snow falls on them. The snow slowly turns to ice when enough pressure has accumulated from fresher overlying snow. Because the enormous pressure of the overlying material is also squeezing the ice, the shields grow sideways along their edges. When the edge reaches a steep coastline, as happens in Greenland, the overhanging ice breaks off and plunges into the sea; this is called *calving,* the splintering chunks of ice forming the infamous icebergs. When the land drops gently into the sea, however, the ice-shield stretches out into the water and floats when it looses contact with the ground. This happens in several regions of Antarctica. The areas consisting of floating ice-sheets are known as *ice shelves.*

At the end of the last ice age, the ice shields that covered Scandinavia, northern Russia, Siberia, and northern North America unloaded into the Arctic Ocean; the north European ice shield passed into the North Sea and the eastern part of the North Atlantic Ocean; and the shield covering Canada and the northeastern United States, the Laurentian Ice Shield, slid into the sea off Labrador, the western North Atlantic Ocean.

As the ice sheets crawl over land, they pick up all kinds of debris, from sand and dust to rocks of all sizes. Some of the load winds up as extra freight on or in the icebergs and is thus carried out to sea. When the icebergs disintegrate, usually from melting, the solid detritus sinks to the bottom of the sea where it merges into the layers of sediments forming there. The several distinct ice ages gave rise to epochs during which such debris was added to the sediments below. These sediments keep in chronological order records of iceberg activity during the epochs.

In a number of places in the North Atlantic, holes have been drilled into the sediments and the cores brought to the surface for analysis. Because the core material matched the rocks common to eastern Canada and the northeastern United States, it was clear that the icebergs from the Laurentian Ice Sheet had served as a means of transport. Six such transportation events, known as *Heinrich events,* have been identified; each is a library documenting the history of the region. The amount of material found in each Heinrich layer and the distance from the coast to which each layer extends allow conclusions about the length and severity of each of the glacial periods identified in this form.

Other means make for precise timing of the occurrences described above. The glacial periods timed in this form coincide with those found by other means. Also, considerably fewer traces of marine life were found in the Heinrich layers than were found in the adjacent intervening layers. During the periods showing greater deposits of debris, reduced productivity of living organisms was noticeable. Heinrich events are also found in South Atlantic sediments, but their timing reveals a displacement with respect to the North Atlantic events by either –3,000 or by +7,000 years. The ambiguity of the displacement's sign cannot yet be resolved nor can any reason yet be given for it.

A wealth of information about the climate of the past can be extracted from the ice sheets accumulated over long periods of time in the two polar regions. Glaciers in high mountain ranges, the Himalayas and the Andes in particular, may cover the history of a few tens of thousands of years. But in the Arctic and Antarctic, precipitation falling only as snow never melts, although a minor loss can occur due to sublimation. The snow, together with the dust carried over by the winds, piles up year after year, creating ice sheets as much as 4 kilometers (2.5 miles) thick.

The weight of the overlying snow compresses the deeper layers and converts them into solid ice. An ice core, like a core of ocean sediments, yields an exact chronological order of events. Annual layers can be readily recognized in the manner of tree rings. Thus dating is a straightforward matter. At the Russian Vostok Camp in the Antarctic, an ice core of 2,100 meters (6,900 feet) had about 160,000 annual layers, the deepest of which represented snow, dust, and air deposited 160,000 years ago. This drilling was later extended, with occasional interruptions, to 3,350 meters (11,000 feet).The last and therefore deepest and oldest ice sample was calculated to be 426,000 years old. The analysis of the first part of the drilling project (the sample that reached 2,100 meters (6,900 feet) deep and an age of 160,000 years) is already nearly complete.

The ambient temperature, or, more exactly, a mixture of the temperature at the ocean surface where the water evaporates and of the cloud level where the snow flakes form, was determined for each layer, using the relative abundances of the hydrogen and oxygen isotopes. The result is shown in the upper curve of Figure 9.4. The end of the last ice age is readily recognized and is in good agreement with previous determinations by other

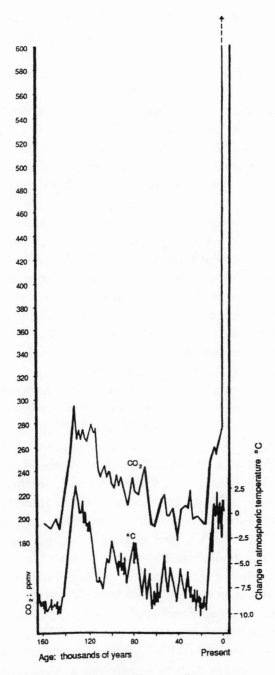

FIGURE 9.4 Two curves showing the close correlation between atmospheric temperature (lower line) and abundance of carbon dioxide (upper line) over the last 160,000 years. Note the predicted increase in the near future.

methods. The curve shows a range in temperature of about 10°C. This does not necessarily mean that the worldwide temperature fluctuated that much; theoretical considerations derived from General Circulation Models indicate that global temperature variations brought about by variations in the greenhouse effect will be maximized near the poles. In tropical regions, only about one third of that amplitude is expected. No similar ice cores are available in the Tropics to verify this conclusion empirically.

Composition analysis of the air in bubbles yields the variation in relative abundance of principal greenhouse gases during the period covered by the core. Of particular interest is the behavior of carbon dioxide, which, along with water vapor, is the most important greenhouse gas in the Earth's atmosphere. The variations in carbon dioxide's abundance are shown in the lower curve of Figure 9.4. The methane content of the air samples was also analyzed, and its behavior is similar to that of carbon dioxide.

A striking resemblance occurs between the two curves in Figure 9.4, although they differ in the details; it is tempting to conclude that the greenhouse gases regulate the global temperature of our planet. This, however, is not necessarily the case, as the following considerations will show. Let us first suppose that carbon dioxide and methane are released into the air from a yet unknown source. The consequence is a warming of the atmosphere through the greenhouse effect. Subsequently, the oceans warm up as well, though with a considerable delay because of their large heat capacity. Warmer oceans lead to an additional gas release from the water and enhances the effect. Also, more water evaporates and further intensifies the greenhouse effect because water vapor effectively absorbs infrared radiation. We have here what is called a *positive feedback*. But things could work out the other way around. Let us suppose that by some not-yet-understood process the oceans warm up first. As a consequence, they release carbon dioxide into the air, thus increasing carbon dioxide's relative abundance in the atmosphere. The positive feedback then works the same way as before: A source of the heat that works first on the oceans and subsequently on the air, in this case almost instantly, could be no more than a rearrangement of ocean currents in the sense that less communication occurs between the warm surface waters and the cold deep layers. Another possibility is that heat was released directly from the interior of the Earth through volcanic and neotectonic activity such as is common to hot spots in Hawaii, the Galapagos Islands, and the Kamchatka Peninsula. *Neotechtonics* is a term for motions of the lithosphere, the outer solid layers that include the crust.

We should remind readers of a paramount difference between land and water. Land is opaque to sunlight and is a good insulator of solar heat. Water is relatively transparent and a poor insulator. At its most intense, sunlight can penetrate ocean water to a depth of 300 to 400 meters (1,000 feet to 1,300 feet) under ideal conditions; only below that level does total and eternal darkness prevail. For this reason, land heats up more quickly than water and cools off more quickly as well. In caves only a few meters underground, the temperature is constant and usually at the average level for the year for their locations. Air quickly takes on the properties of the underlying surface. Over land, the daily temperature range of air is perhaps twice as great as that over the sea; this fact has many ramifications in climate. Marine climates are less extreme than their continental counterparts, and the lag of the seasons is greater along the coasts than it is inland. The onshore and offshore breezes of a summer day result directly from the difference. In the daytime, air over the shore heats up and expands, causing it to rise; air over the sea flows in to replace it. At night the process reverses because the air over water cools off more slowly. The entire process is called a *simple heat engine.*

The analysis of the remaining ice cores obtained in the Vostok project is incomplete as this book goes to press. Nevertheless, data on the abundance of hydrogen and oxygen isotopes, proxies for the ambient temperature, are already available. Four interglacial periods are clearly visible, and their timing coincides with previously known dates.

Ice-core drilling has also been carried out on three sites in Greenland, of which one is at the summit of the ice sheet, and the others are near the coast. The first two Greenland drilling projects, one started in 1992 by European scientists, the other in 1993 by Americans, are only about 30 kilometers (about 20 miles) apart. Excellent agreement has been found in the climate history from both experiments, with the exception of a 12,000-year interval occurring a little more than 100,000 years ago where the two records sharply disagree. Was there a mix up of the ice samples as they were brought up, or is the disagreement due to the uneven surface of the bedrock under the ice? The site of a third project is 340 kilometers (about 200 miles) from the other two sites; radar tests show that at this site the rock surface is smooth, which may resolve the controversy.

Through advances in techniques, the Greenland ice cores, in particular the core from the Summit site, have been studied with considerably higher resolution than has the Vostok ice core. The Summit core drilling reached solid rock at a depth of a little more than 3,000 meters (about 10,000 feet), and is expected to cover the climate history of the last 200,000 years. One notable result of this information is that volcanic dust in the ice sample provided a new dating for the great volcanic eruption that all but leveled the Aegean island of Santorini, also known as Thera. It is now widely accepted that this event gave rise to the legend of Atlantis, described by Plato. The eruption happened around 1630 B.C., earlier by a century or two than dates Plato suggested. This catastrophe had a great impact on human affairs; the ash and dust from the explosion covered much of the island of Crete and destroyed its agriculture. Allegedly, the disaster weakened the Minoan civilization and enabled the more barbaric Mycenians to the north to invade and conquer.

The amount of ice extracted in core samples is so enormous and the process of analysis so cumbersome that many of the results are not in yet. Scientists give particular emphasis to a high-time-resolution study of the last 40,000 years, now complete, which detects just how fast significant changes may occur in the polar climate.

The temperature analysis of the Greenland ice cores yielded surprising results. Climate variations of more than 5°C (9°F) did occur several times in very short time spans, occasionally even within a decade. Scientists believe that only a sudden change of the trajectory of the ocean currents can cause such a rapid climate change. The air temperature above the ocean, and therefore the climate, is strongly influenced, almost dominated, by the surface temperature of the water. Furthermore, the air adapts almost immediately to changes in the surface water temperature. The opposite, naturally, does not happen; water takes a long time to respond to temperature changes in the air. (In Chapter 7, we explained in detail the mechanism of ocean currents and their linkage to the atmosphere.)

Glaciers present a number of extra problems when compared to the extensive ice sheets on Greenland or in Antarctica. Because they are often moving relatively fast, the ice can expand horizontally, particularly at greater depth and therefore greater pressure; this can lead to such a thinning of the individual yearly layers that timing through layer counting becomes impossible. A cruder timing is still possible with the help of radioactive materials and their abundance.

Suitable drilling sites have been found on the high plateau between Tibet and China, where an ice record goes back 500,000 years. The climate history there agrees quite well with those of Antarctica and Greenland and shows that climate variations were global and simultaneous in both hemispheres in the Tropics. Mt. Huascaran in Peru, at an elevation of over 6,000 meters (20,000 feet) was another promising site in the Tropics. Its record goes back 12,000 years, far enough to catch the end of the last ice age. It also shows, in good agreement with all other historical climate records, a particularly warm period from about 8,000 to about 5,000 years ago. Today, it is again as warm as it was during that period. The total temperature variation at the elevation of Huascaran is about 12°C (22°F).

Not only can the temperature of the atmosphere be calculated for large time intervals from ice cores. As we have already mentioned, the enclosed air bubbles contain a historical record of the composition of the air. The abundance of carbon dioxide as the major climate factor in the regulation of the greenhouse gases is particularly interesting. Gases such as methane and nitrogen oxide may hold clues on the plant-growing activity at the respective times. The aerosols embedded in the ice also have a story to tell: A high component of dust seems to indicate a dry year and accumulations of volcanic ashes mark the times of major eruptions. The timing of the legendary Santorini eruption of 1630 B.C. was confirmed through the study of aerosols; as we noted above, this eruption accounted for the legend of Atlantis.

Permanent ice sheets are also found in sub-polar regions. When some melting occurs during the summer, water then penetrates some of the ice layers below it, giving them a characteristic aspect. An analysis of melting layers in Canadian ice sheets showed that the warmest summers occurred 8,000 to 10,000 years ago, shortly after the end of the last ice age. The records of these ice sheets do not go farther back in time. Surprisingly, the coldest summers indicated by the melt layers occurred only 150 years ago.

Sediments forming at the bottom of a lake can preserve remains of organic matter in a chronological order providing no disturbances occur. Pollen of most types of plants, recognizable even when fossilized, is surprisingly durable even in the most adverse conditions. Sediments in lake beds provide a detailed account of pollen production for various species of plants throughout the history of the lake. Pollen production depends on the climate, particularly on its temperature and moisture.

Lakes in a closed basin can yield historical climate information through another mechanism. The water levels of such isolated lakes are determined by two factors: the annual influx of water, which is a consequence of local annual precipitation, and the evaporation rate, which depends on the temperature and humidity of the air. In this way, lake levels vary in accordance with local climates. In many of these lakes, particularly in Africa, water levels during the past few tens of thousands of years have been reconstructed from their varying shorelines.

Analysis of the relative abundances of stable isotopes (atoms of the same element but with different numbers of neutrons—chargeless particles—in their nuclei, but not in the atomic charge) in the remains of organic matter yields direct information on the ambient temperature at the time the organisms lived. This type of analysis can be carried out in any sediment, be it from a lake bed or from the bottom of the ocean; for example, oxygen isotope temperatures from a sediment in Nevada covering 320,000 years yielded results in good agreement with the Vostok data.

Another method, based on distribution of plankton fossils, reconstructs the history of the oceans' surface temperatures. Marine plankton are sensitive not only to the temperature but to the chemical composition of the water near the surface, the region they inhabit. Because some plankton prefer warm waters and others cool waters, the distribution of plankton species varies with the water temperature; when the plankton die and sink to the bottom like so much ballast, their distribution, and thus a temperature record of the surface, is preserved through the locations of the fossils. Plankton, therefore, are one of the important contributors to the forming sediments.

These studies and others have been conducted on a grand scale through the combined efforts of many investigators as they combed sites all over the world. The records of the project cover almost 20,000 years. The main purpose of this joint effort was to obtain simultaneous data from all over the world for comparison with the General Circulation Models. Only when the models satisfactorily explain the past can we rely on their predictions for the future.

For the most part, investigations reveal that climate variations, particularly in temperature, were coincident over the entire planet, although wet conditions in one part of the world may have led to dry conditions in another. Studies of ocean corals show that they were submerged to a very low level in the Caribbean Sea at the same time they were low in the Far East. Glaciation around the North Pole coincides with glacier growth in the Andes, as do terrestrial and marine temperature variations from southeast Africa. These are just a few of the existing examples, and they contain differences in the details. For example, investigations in France show that in one region the maximum extent of the ice sheet was reached during one glaciation period, but in another region the maximum occurred during a different period.

From Vostok ice core data we see that the glaciation periods coincide with minima in the abundance of carbon dioxide in the atmosphere; this raises the question of the missing carbon, discussed in Chapter 5. Part of it could have been stored in the vegetation on land, but not all, as theoretical calculations have deduced. The oceans are more than capable of absorbing the requisite amount of carbon dioxide. Not only can the water have dissolved the missing gas, but even more so can an increased plankton growth have caused a more permanent extraction of carbon dioxide from the atmosphere. Some have even suggested that plankton regulate the carbon dioxide content of the atmosphere and therefore the climate.

The data on hand have been used to determine whether climate fluctuations are periodic. In particular, a search was made of the periods of 19,000; 23,000; 41,000; and 100,000 years predicted by Milankovic. We discussed the astronomical bases for these periods in Chapter 4, and all of them turn out to be present in the climate variations; they do not, however, explain all the variations. Other sources, periodic or nonperiodic, must be involved. We are now aware of another source that enters the picture. Whereas all past climate variations, whether terrestrial or extraterrestrial, originated naturally, we must account for manmade alterations of the environment.

10

THE POST-GLACIAL
PERIOD: HOMO SAPIENS
COMES OF AGE

THE HOLOCENE PERIOD, which began about 10,000 years ago, is our period, our time to dominate the world and its other denizens. We have not done very well by them in the past, but some encouraging signs tell us that we are aware of this and might reform. Wanton slaughter of mammals, at least, has become properly taboo in many parts of the world. The Holocene became the period of warmer temperatures as the last ice age finally receded (see Figure 10.1).

Apart from the difficulties of piecing together our weather history, outlined in the last chapter, incomplete and nonuniform coverage of the Earth by cloud observers raises more problems. The eastern United States has been well covered since the beginning of the eighteenth century, and the same is true for western and central Europe. Few data are available from other parts of the world, with practically none in the Tropics or over the oceans. Data taken at sea by ships' captains or travelers are scarce and nonuniform. Reliable data with a nearly worldwide coverage can be found only for the last century or less, and even then, mostly over land.

Several investigators have collected and sorted temperature and precipitation data from as many sources as they could uncover the world over; they had to discard most of the data for any of the reasons explained earlier. It is not surprising, then, that investigators arrive at different results. In Figure 10.2, we show one of the latest temperature history curves. A trend of rising temperature is certainly visible, but the amplitude of the trend lies within the uncertainties inherent in the data and compilation

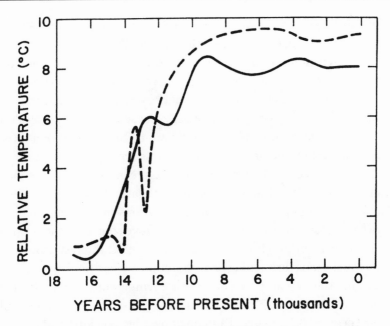

FIGURE 10.1   Average temperatures in both hemispheres over the last 17,000 years from ice-core data.

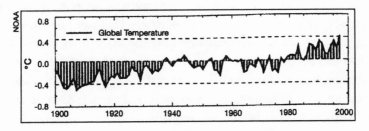

FIGURE 10.2   Combined land and sea temperatures over the last 100 years relative to the 1961–1990 average.

procedures. One can tentatively conclude that during the last one hundred years, global temperature has increased about one degree Celsius.

It is tempting to attribute the rising trend to the accumulation of carbon dioxide and other greenhouse gases in the atmosphere, released into the air by the burning of fossil fuels and other human activities. Certainly the two are correlated, but correlation is not necessarily causation. In the early 1990s, The Intergovernmental Panel on Climate Change (IPCC) found evidence for a "discernible human influence on global climate." We must admit, however, that nature has its own ways of altering the global climate, sometimes on a much larger scale, as we have shown in previous chapters. We need more data over a longer period as well as a solid theoretical understanding before we can definitively answer this question; certainly, satellites and computational sophistication show promise here. Action on global warming should not depend on the final placement of every nail that supports a theory; we should begin where we cannot deny plausible and reasonable doubt. Our somewhat heuristic viewpoint would not be the first case of a theory's being fully accepted before final and incontrovertible proof was in: The Copernican heliocentric theory was fully accepted a century or two before its confirmation. We will address this point in a later chapter.

The IPCC was established by about 30 nations in 1988, and scientists everywhere recognize it as the most authoritative source on the science of climate change, although some cranks and special interest groups who find its conclusions threatening deny its authority. The IPCC's 1990 report on warming shows that the average surface temperature around the globe has increased 0.80° ± 0.27°F (0.5°C); that is, eight-tenths of a degree with a standard error of 0.27 degrees Fahrenheit over the previous century. Later, in 1995, it reported that around half of this warming had occurred in the last 40 years, a view contrary to the assertion that most of this warming happened before 1940. The increased temperature was not uniform, but was greatest over the continents in the Northern Hemisphere. The diurnal range in temperature has decreased because the nighttimes have warmed more than the daytimes, correlated with an increase in cloud cover. Furthermore, the last year for which data are complete, 1998, was the warmest on record, and the 10 hottest years have occurred since 1983, at least since reliable weather records began. We will discuss these and other findings more fully in Chapter 17.

Other methods of obtaining information about the recent history of the climate include the ever-popular study of tree rings. At intermediate and high latitudes trees have a specific growing season and form annual rings in their trunks accordingly. The width of these rings is related to the spring and summer climate, particularly temperature and precipitation. If conditions favor growth, the rings are wide; unfavorable conditions lead to narrow rings. Interpreting the ring widths can be ambiguous because they depend on both temperature and precipitation, and possibly the availability of nutrients. This relationship can be calibrated wherever simultaneous ring widths and climate data are available. In most cases, the temperature has greatest influence on the widths of the rings.

The first studies on tree rings were made to determine whether a relationship exists between the periodic sunspot activity and the climate. The results showed that the solar cycle of 11 years is involved, although inconsistently, indicating that other effects could also influence the climate. We will discuss these in the next chapter. If we accept that the widths of the tree rings are indicative of the temperature—the mean temperature during a certain part of the year—any tree trunk or stump, dead or alive, can tell the story of the climate for the span of its life. This raises a question: Do tree rings reveal a climate of continental significance, or do they represent local climatic variations only? This problem was studied in much of western Europe and covered the years from 1850 to 1971; the study proved that a close correspondence between the patterns of tree-ring growth extended over large continental regions.

The oldest tree-ring samples come from trunks several thousand years old and preserved in swamps. Each tree covers only a restricted time lapse, but such distinct details as several successive favorable or unfavorable years can help us assemble the pieces of time into one continuous record. Today, the climate history from tree rings covers more than 6,000 years, although a number of large gaps are yet unfilled. Naturally, from carbon isotope analysis (isotope decay being an indicator of age) one knows at least the approximate timing of each of the pieces. A continuous record from about 1,000 years ago all the way to the present comes from Australia's island state of Tasmania, where a forest so far undisturbed by man supplied the required dead and living trees. The data exhibit a semiperiodic fluctuation with a period of about 20 years and a substantial temperature increase during the last one hundred years, in good agreement with the temperature data mentioned above (see Figure 10.3).

Deposits made by rivers in their estuaries may reflect their annual flows; these depend on the annual precipitation. A study of the Amazon River's annual flows covers the years from 1903 to 1985. No long-term trend is detected, but there appears to be a correlation with the so-called *El Niño period*, a phenomenon we will examine more closely in a later chapter.

The existence and growth of most types of vegetation are sensitive to the climate. A given species grows only within certain limits of latitude and elevation. For trees, these limits or boundaries are known as *tree-lines*. When the global climate changes, climate zones shift north or south as well as uphill or downhill; the tree-lines, and of course any lines referring to other species, migrate along with the climate zones. Tree-lines can be recon-

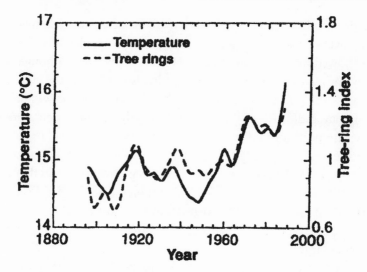

FIGURE 10.3   Temperatures and tree ring indices for Tasmania.

structed from historical as well as from geological evidence; this intro-
duces for study a new feature related to the climate.

The mechanism that makes the trees and other plants follow the shift of
the climate zones is quite evident: Plants along one border of the zone they
inhabit may find themselves in an unfavorable climate and no longer able
to prosper, yet along the other border they find a new and suitable area;
this area in turn is being abandoned by the plants that existed there before.
Because such migration concerns not one species only but an entire eco-
logical community, the settling of new territory is a slow process. From
historical as well as from geological evidence it was found that, depending
on the species, these propagation rates vary from about 1 to 45 kilometers
(0.62 to 28 miles) per century. The rate is determined by the spreading
range of the seeds of the entire community and raises an interesting and
important question: What happens when the climate zones move at a
faster rate than the propagation rate of the species involved? One can
imagine that some species disappear. The disappearance of species in this
way is of great concern for the future because the present migration rate of
the climate zones is from ten to one hundred times faster than it was during the past hundreds of thousands of years.

In temperate zones, the single most significant climate constraint on
plants is winter hardiness, defined by the average coldest temperature
reached during a winter. Nurseries stock only those trees and shrubs that

Range average annual minimum temperatures for each zone

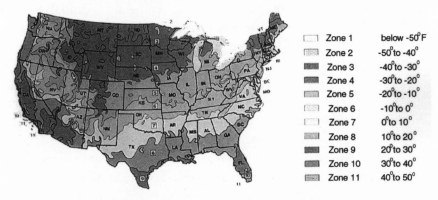

FIGURE 10.4   USDA Plant Hardiness Zone Map. The zones are delineated by average minimum temperatures.

can survive this coldest temperature in their marketing area. Winter hardiness zones in the United States and Canada are defined such that each zone covers a temperature range of 10°F (about 6°C). Zone 6 is defined as the area with coldest temperatures, usually falling between 0°F (-18°C) and -10°F (about -23°C). A glance at the United States Department of Agriculture (USDA) Hardiness Zone Map (see Figure 10.4) shows that most of Connecticut lies in zone 6; weather records for Connecticut show the temperature falls below –10°F (-23°C) in fewer than one winter in every four. This zone is fortunate in that it lies within the southern limit to the range of the sugar maple, which, more than any other tree, gives New England its glorious autumnal color. Yet the state maintains many flowering trees and shrubs that will not flourish to its north.

The American sweet gum tree, known also by its Linnaean—or botanical—name, *Liquidambar styraciflua,* serves as an example. *L. styraciflua* is a well-known specimen tree with superb long-lasting fall foliage. Although not indigenous north of New York City, it grows wild in Maryland and southern New Jersey (lying in zone 7) and points south. In Connecticut, in zone 6, where rainfall is usually abundant, *L. styraciflua* grows reliably and well and is mostly unaffected by the air pollution found along roadways and in urban areas. The sweet gum is not winter hardy in zone 5, just north of Connecticut, and is not found there. In their turn, many varieties of azalea and rhododendron, as well as *Cornus florida,* the flowering dogwood tree, flourish in zone 5 (upstate New York, western Massachusetts, southern Michigan) but not in zone 4 (Minnesota, northern Maine). Ten zones in all range from the first, where the temperature can fall below −50°F (-45°C), to the tenth, where temperatures rarely fall below freezing. Zone 10 may serve as the definition of the limit to a tropical climate, where a killing frost never occurs.

The burnished appearance of the sweet gum's foliage betokens a tree that hails from lush early prehistoric times. Its three surviving species are widely spaced around the world—only three of more than twenty species that were alive a few million years ago. The Earth was warmer and milder then, before the start of the Pleistocene epoch: Its ice ages occasionally scoured and devastated much of North America and Eurasia and left them under thick sheets of glaciation, in the process making extinct most of the species.

Other factors are also important to the adaptability of a plant to a particular environment; these include the extent and frequency of drought, the acidity of the soil, the elevation above sea level, and the presence of pollution or blight. Any long-term change in climate also affects the presence or absence of poor growing conditions. For example, the increasingly severe cold snaps and frosts in central Florida have caused the citrus belt there to move southward by the width of a county, about 20 miles (30 kilometers).

In localities where frosts are rare or absent altogether, elevation is the most significant player in plant survivability. The giant saguaro cactus flourishes near Tucson, Arizona (or did until most were ripped off by poachers) at an elevation of 2,400 feet (750 meters), but not at nearby Las Cruces, New Mexico, at 3,900 feet (1,200 meters) at the same latitude and not far higher. In Mérida, Venezuela, stands the longest cable car in the world. The first of four stages begins in Mérida at about 1,600 meters, a mile-high city like Denver, and the fourth stage tops off at just under 5,000 meters (16,000 feet). Just 9 degrees of latitude north of the Equator, the summit of the mountain boasts glaciers not seen at sea level much below the Arctic Circle, whereas the city below thrives in a tropical climate. In North Carolina, the weather station atop Mount Mitchell, at 6,684 feet (over 2 kilometers) above sea level and the highest point in eastern North America, reports a winter temperature and snowfall equivalent in severity to that of Minneapolis, far to the north.

•   •   •

Undisturbed soil can also yield the recent temperature history. The first few centimeters of depth of the soil respond to the diurnal temperature variations, the insolation, and the precipitation. Seasonal variations leave their temperature signals only a few meters deep. But long-term temperature variations penetrate to greater depths: The temperature profiles measured in undisturbed soil down to about two hundred meters can reveal temperature trends during the last several centuries.

The cultivation of grapes for wine production is dependent on the climate. One finds in Europe a well-defined northern boundary to the grape-growing areas. This boundary has shifted considerably north and south during historical periods. In Roman times, grapes were raised throughout the entire region of Germany occupied by the Romans. Nowadays, only south-facing slopes are suitable for growing grapes, and at times practically no wine is produced in Germany. In Britain, too, vineyards were plentiful in medieval times. Southern England had a number of them active from 1000 to 1300, when summer temperatures averaged almost 1°C (34°F) warmer in England and nearly 1.5°C (35°F) in central Europe than they do in the twentieth century. The extensive and luxurious Roman open-air baths at Trier, Germany, and Bath, England, also indicate that the climate then was milder than it was over most of the twentieth century. Indeed, various records of the success or failure of other agricultural activities point to substantial and long lasting climate changes.

A noteworthy event was studied by H. Stommel and E. Stommel in their book, *Volcano Weather*. The authors refer to 1816 as the famous "year without a summer." During that year, the summer in the eastern United States was unusually cold, with frosts throughout the summer and snowfalls recorded in New England in June and July. Almost all crops were destroyed by the unusual weather. Legend has it that the *Old Farmer's Almanac* made its reputation when a typographical error predicted snow in July 1816. But the journal did not predict snow or frost at that time, as can be seen in Figure 10.5.

During that same year, 1816, similar cold weather, though not quite as drastic as that in New England, prevailed in central and western Europe, where the direct effects of the bad season coincided with the turmoils at the end of the Napoleonic War. Other parts of the world, for example, South America or the Far East, were not affected at all. The authors attribute the abnormal meteorological situation to the eruption of Mt. Tambora, in Indonesia in April 1815; the eruption occurred just two months before the Battle of Waterloo, the battle that ended Napoleon's career. This eruption was the most severe eruption of any volcano during the last 10,000 years. The enormous amount of ash, dust, and gases ejected

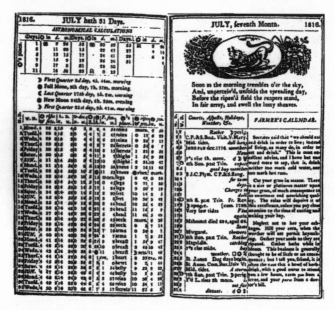

FIGURE 10.5   Entry in the *Old Farmer's Almanac* for July 1816. Contrary to legend, no predictions of unseasonable frosts were made.

into the atmosphere was more than enough to affect the climate, although at that time no one could relate the peculiar weather to the eruption. (Benjamin Franklin suggested in 1783 that a spell of unusual weather during that year might have had something to do with a volcanic eruption in Japan a few months earlier. Dr. Franklin could not have known of the Tambora event because he died in 1790.)

The Tambora eruption of 1815 ejected some 100 cubic kilometers, or 25 cubic miles, of volcanic debris into the atmosphere, where it shrouded the entire globe. This is three to five times the amount ejected during the next three most severe volcanic events in recorded history. The loss of life, estimated at 92,000, also stands as a record for a volcanic eruption. We have referred to one of the next three, the explosion of about 1630 B.C. on the Greek island of Santorini (Thera) in the Aegean Sea, and responsible for the Atlantis legend. The others are Krakatoa, a small island between Java and Sumatra, which killed 36,000 people in 1883, and Katmai in Alaska in 1912. The prehistoric eruption of Mazama in 4600 B.C. was almost as destructive as Tambora; this event created Crater Lake, now a national park in Oregon.

Tambora exceeds other well-known eruptions by a factor of almost a hundred, including the infamous eruption of Vesuvius in A.D. 79 that destroyed the Roman cities of Pompeii and Herculaneum. Two others about on a par with Vesuvius are Mount Pelee on the Caribbean island of Martinique in 1902, killing 30,000 people, including all but two in the town of Saint Pierre, and Mount Saint Helens in Washington state in 1980. The most recent volcanic event of this type was that of Mount Pinatubo in the Philippine Islands in the summer of 1991. Its ejections colored sunrises and sunsets (and the risings and settings of the full Moon as well) for about a year afterward. A white haze hovered around the Sun and the full Moon even on the clearest of days, and fewer visible stars shone than in normal times. Another eruption, that of Mount Agung on the island of Bali in 1963, also affected the weather and the sky.

The effect of volcanism on the climate is only one part of the picture. Ice core records from Greenland and Antarctica document global climate and volcanic activity for the last 500,000 years. The records showed that increased volcanic activity is present whenever the climate undergoes changes, whether they be cooling or warming. Sea-level changes that result from the forming or melting of large masses of ice and the changes of ice shield loads on continents create a changing pressure on the magma under the lithosphere, and may therefore trigger eruptions.

Within the uncertainties of all methods involved, a rising trend in the global temperature during the last one hundred years emerges. Naturally, it would be interesting to know whether a trend also exists in other climatic parameters. Some historical data for the precipitation on particular sites are available, one of which is shown in Figure 10.6. No global curve of this kind can be constructed without data, and as we have mentioned already, this is also true for the global cloudiness.

In some specific times and places we have a clear idea of weather conditions. Eyewitnesses have provided descriptions of weather conditions at Hastings on October 14, 1066; in Philadelphia on July 4, 1776; and in Dublin on Bloomsday, June 16, 1904. Even at latitude 41°46' north and longitude 50°14' west in the North Atlantic Ocean on April 15, 1912, and at Pompeii on August 24, A.D. 79, the weather conditions are well documented. These are informative for the events they surrounded, but they are of little value for an intimate knowledge of weather conditions in general and of how they varied from the norm for earlier or later times.

The concept of the ice age as evenly cold and the subsequent Holocene Epoch of the last 10,000 years as evenly warm is simplistic and does not

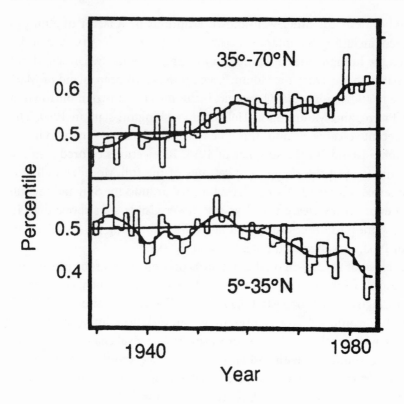

FIGURE 10.6   Rise of rainfall in Europe (top line) and decline of rainfall in Africa (bottom line).

represent the truth. The Wisconsin period of glaciation had a warmer interstadial period about 25,000–30,000 years ago, and a sharply cold minimum 18,000 years in the past. Similarly, the temperatures of the Holocene period have also varied, turning alternately warmer and colder in times past. We have already shown how these differentiations in temperature are discovered, and to that we can add the evidence that historical documents have given us.

Ocean levels relate closely to the global temperature average. Thus, when we find Roman piers and docks built at a level higher than would be realistic today, can we not conclude that the Romans lived in a warmer era than our own? In Roman times, the seas were higher because the ice caps had melted more ice into them. Similar effects can result from the motions of tectonic plates making up the Earth's crust, which give rise to continental drift.

From human artifacts and records we can define four periods of more than passing interest since the great retreat of the glaciers over 10,000 years ago. The first of these lasted from about 5000 to 3000 B.C. and is known as the *postglacial climatic optimum* or the warmest period of the Holocene. (Equating optimal conditions with warmest average temperatures is quite arbitrary, and this term is being abandoned as a result.) The warmth coincided with the rise of river civilizations along the Nile, Indus, and Hwang Ho Rivers among others. Agriculture flourished and permitted the rise of city-states, and the mild conditions may have helped secure their relative permanence. Ocean levels were about 10 feet (3 meters) above those of today, and sea ice around the Arctic Ocean had shrunk to a minimum, indeed to such a point that Greenland and the Arctic Archipelago of northern Canada were mostly surrounded by open water.

Then came a colder climatic epoch; this culminated between 1000 and 300 B.C. The rise of Greece happened during this time, but northerly lands did not thrive; perhaps they were too cold for much development. Historical records of classical times play an important role in the definition and identification of the weather of the time.

A secondary climatic optimum rising to a temperature maximum in the high Middle Ages of A.D. 1000–1200 followed the colder climatic epoch. The melting of pack ice in the North Atlantic led to the establishment of colonies along the island stepping stones that stretch across the far north from Europe to North America. Vineyards, a reliable indicator, again spread north some 4 degrees of latitude and from 100 to 200 meters (about 328 to 656 feet) in altitude along mountain slopes. Timber lines also climbed upward as glaciers retreated to the heights.

Sometime towards the end of the first millennium, the Nordic races of Scandinavia developed their amazing longboats. About A.D. 1000, when the Viking incursions had ended, a period of exploration and settlement began. The Norsemen discovered and settled southern Greenland. Agriculture there flourished to the extent that a number of communities arose along the coasts. By about 1200, Greenland had its own bishopric with seventeen churches and several monasteries; seafaring and trade flourished.

The exploits of Eric the Red and his son, Leif Ericsson, are well known. How far they traveled in their explorations is still open to conjecture, but we do know that shortly after the year 1000, Leif Ericsson and his men sailed beyond Greenland to a place they called Vinland. It is now accepted that these people got at least as far as the northern tip of Newfoundland;

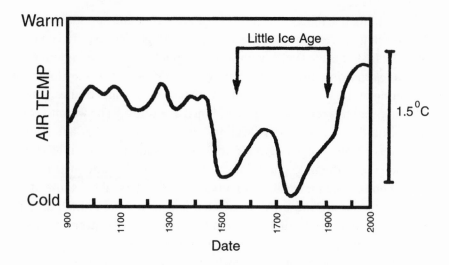

FIGURE 10.7   The variation in average temperature over the last 1,000 years. The colder period of the Little Ice Age is clearly shown.

we have confirmed that a Norse settlement existed at l'Anse aux Meadows, where ruins are still visible.

Then came the last and most studied of the four, known as the *Little Ice Age*. Climatic changes occur on several different time scales. Some last for a few years, others for thousands or even tens of thousands of years. In between are variations of a few centuries; the most widely studied of these periods happened over the last several centuries, mainly over northerly latitudes. This Little Ice Age began in the thirteenth and fourteenth centuries and culminated in the seventeenth and eighteenth centuries, finally ending just over a century ago. As its name implies, the Little Ice Age brought low temperatures to Europe and North America, and its effects were felt worldwide. In the nineteenth century, its cold extremes abated, and more "normal" temperatures returned. The mean temperature for the past 1,000 years appears in Figure 10.7. The effect of the Little Ice Age is evident.

The Little Ice Age is important to us for two principal reasons: It occurred during a time of substantial civilization over much of the Northern Hemisphere. Many similar climatic variations have befallen earlier times, some of them severe. But as any study about the influence of changes in climate shows, the effect upon civilization is much greater than upon our species in the large. People in general can easily survive a major weather event; the powerful El Niño has not affected overall populations, disastrous as it has been upon certain small regions. An asteroid such as the one that formed the Barringer crater in Arizona would not come close to destroying a civilization, but it would demolish any city it crashed into. Even a recurrence of the K/T disaster, which killed off the dinosaurs, just might not wipe out Homo sapiens, spread as we are across the globe; but not a speck of what might be called civilization could hope to survive. As societies spread and grow more complex, they are more vulnerable to major disasters. In this vein, the Little Ice Age affected us more than did any historical equivalent.

The second reason for intensive study of the Little Ice Age is its recency; it is thought to have waned over the late nineteenth and early twentieth centuries, but we do not have a secure baseline to tell us with certainty that its influence is over. We know that global temperatures rose over the period beginning in 1850 or 1880 until 1940; then a downturn appeared to last from 1940 until near 1970, with a subsequent rise after 1970.

Just what happened is now clear, even if the causes are not. The Little Ice Age has been documented by more methods than might be imagined. Some are dubious by themselves, but together they confirm the major climatic irregularities of the period.

Today, anyone whose transatlantic flight has been diverted northward over the wastelands of Greenland and Labrador sees firsthand that they are inhospitable to any but the sketchiest human existence. Yet ever since the end of the most recent ice age, our climate has vacillated between periods of relative warmth and coolness, often lasting for centuries. The recent warm period that began about the time of the rise of Hellenistic Greece and the Roman Empire lasted well into the late Middle Ages with some variation along the way. During the last part of this period, the lands of the North Atlantic flourished to a degree beyond anything seen since.

Then in the fourteenth century, as we have mentioned, something happened to the weather. For reasons we still do not fully understand, the climate of Greenland cooled markedly. The Gulf Stream diverted to the south and Greenland and Iceland suffered the results. When the harbors became icebound, trade and agriculture halted. Contact with Europe was severed and the European inhabitants of Greenland vanished in a manner yet unknown. Perhaps they intermarried with the natives living there, or maybe they simply froze or starved to death. In any event, Greenland and its settlements were lost and forgotten.

About a century later, northwestern Europe followed Greenland into the cold and winters turned much harsher. How do we know this? Although

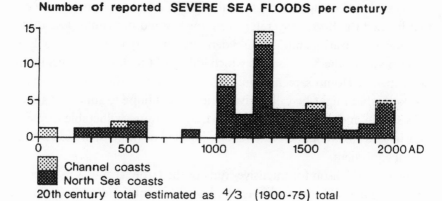

FIGURE 10.8   The distribution of severe floods per century in northern Europe over the last 2,000 years. Note the relative minimum during the medieval warm period.

the thermometer and barometer were not invented until the seventeenth century and therefore maintaining quantitative records was impossible before that time, other observations were rife. The relative warmth of the high Middle Ages after Charlemagne is documented by the growth of tree rings, by the higher locations of timberlines in the Alps, and by numerous contemporary reports of the time. Records show that the most northerly cultivations of grains expanded and that vineyards were common in England, a sign of higher summer temperatures then than now. The age of the Crusades and great cathedral building is testimony to vigorous human activity.

This cold period is delineated also by fewer severe floods along the coasts of Europe. During the three hundred years before 1300, and again in the twentieth century, the number was greater than during the interim (see Figure 10.8). This distribution suggests that storm floods along the low-lying coasts around the eastern shore of the North Sea were more prevalent when the sea level may have risen after warm periods accompanied by glaciers melting.

The Little Ice Age has also made its mark in the art of the times. Who does not know the paintings of Pieter Bruegel the Elder in the mid-sixteenth century? His landscape *Hunters in the Snow* is one of many winter scenes during the cold winters that lasted for two hundred more years. Even John Constable's rural pictures of East Anglia provide clues to changes in the weather. Statistical studies of the cloud cover in European representational-style paintings show a decline in cover from 70 to 80 percent when Constable and Turner painted (about 1790–1840) and before; these studies show the decline in cloud cover reached from 55 to 70 percent early in the twentieth century. Many of Constable's landscapes show billowing cumulus clouds more characteristic of his time than ours. Cloudiness is indicative of cooler weather. Literature and couture also provide supporting evidence for colder times; the white Christmases of London portrayed in the novels of Charles Dickens document a time colder then than now; and clothing styles of postrevolutionary Paris were geared towards colder winters.

In the United States, the single best known event marking the period of cold is the winter of 1777–1778, when George Washington and his men endured severe hardships in the cold of Valley Forge near Philadelphia. Although the following winter was even colder, this is the strongest collective memory of those hard times, and it is an accurate one.

Photography documented the warming years following 1850; we see glaciers in retreat and snow cover on mountains. Records tell us that the Thames River in London froze over completely for the last time in 1898, although it had frequently done so in the years before.

The Maunder Minimum, which refers to the frequency of sunspots and lasted from 1640 to 1715, is also worthy of discussion. Although the Sun may bring about long-term climatic changes through a variation in the solar constant (the amount of energy it emits), the only well-established periodic variation in solar activity is the eleven-year sunspot cycle. In eleven years, the number and size of spots covering the solar surface undergoes a variation from a maximum through minimum and back again to maximum. This cycle appears to be of little influence in temperature or rainfall on the Earth. From 1640 to 1715, however, the Sun showed no spots at all; it was unusually quiet. Telescopes of the time, and even naked-eye accounts of the absence of larger spots, have confirmed this anomaly. Although it occurred in the middle of the Little Ice Age, it seems to have had no part in the weather. Earlier, several other long periods of solar inactivity occupied equivalent periods over the previous thousand years; they also had no apparent effect on the climate of the times.

11

THE GREENHOUSE
GASES: MAKING OUR
OWN WORLD

We pray for one last landing
On the globe that gave us birth;
Let us rest our eyes on fleecy skies
And the cool green hills of earth.

ROBERT HEINLEIN, *GREEN HILLS OF EARTH*

CONCURRENCE BETWEEN TWO DEVELOPMENTS in relation to a growing problem may or may not be commonplace. For example, outdoor lighting has been part of our urban environment for more than a century, but only in the years after World War II did it become excessive. With cities and their suburbs awash with light, most glories of the night sky are hidden to urban dwellers. But now legislation to shield some lights and remove others is underway. So is it with our climate.

Today, the climate (and climatology) is subject to two major truths. The first is that our tinkerings with the atmosphere are more than equal to the changes nature brings. The second is that we are aware of it, but can we correct our excesses? Climatology is one of the fastest growing fields of science, and as a result, it is beginning to carry political clout.

The systematic collection of weather data and records began more or less with Benjamin Franklin in the eighteenth century. He and others realized from observations at a number of places that weather systems usually arrive from the west (the prevailing westerlies of the midlatitudes are responsible for this, as we already know). Because weather stations across the United States and Europe became widespread only after the American Civil War, the records of most stations do not cover much more than a century.

An extensive network of weather stations now in place worldwide supplies accurate weather data for almost instant global evaluation. We can thank the artificial satellites for making the greatest contributions to weather technology; indeed, since the first meteorological satellites were launched, meteorologists have kept continuous and global records of the all-important cloud cover for the entire planet; they can also distinguish cloud types on the satellite pictures and thus make accurate predictions.

As a result of even greater technological advances, other satellites were launched into orbit; these measure the surface temperature, check the chemical composition of the air, and monitor the pollution levels created by particulate matter. Wind speed and direction over the oceans can now be determined from satellites, and soon precipitation will also be measured. New advances in ground-based equipment have been developed to follow atmospheric characteristics over the sea as well as above the land masses. And computers, with their gigabyte capacity, permit all these observations to be processed well before the time of the prediction has passed.

The mass of data now available give us several reasons for expecting a manmade climate alteration, whether now or in the very near future. One of these is the rapid increase of the greenhouse gases, particularly carbon dioxide: Its invasion into the atmosphere from manmade sources comes directly from the burning of fossil fuels. It concentrates in the most populated and industrialized regions of the world, where the amounts may vary from day to day, or even from hour to hour, depending on the rate of burning fuels and the prevailing wind. The vertical distribution of the pollution depends on the temperature profile (i.e., the variation of the air temperature with altitude). If a temperature inversion (a reversal of the normally descending temperature gradient with height) is present, the released gases are likely to hover in the lowest levels of the atmosphere until ascending air fills in the inversion or until the air mass moves away and normal mixing can take over. Temperature inversions create ideal conditions for smog, especially in regions with mountains to the east. This is why western cities throughout the Americas from San Francisco and Los Angeles to Santiago, Chile, are so susceptible to smog.

To obtain a representative record of the atmosphere's carbon dioxide content as well as any evidence of a global warming trend, measurements must be made not only in or near cities but also at sites far from the biggest emission centers, preferably on a high mountain on a remote island. An ideal site is the Mauna Loa Observatory on the island of Hawaii,

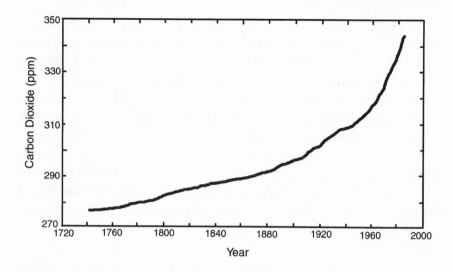

FIGURE 11.1   The abundance of carbon dioxide in the atmosphere since 1750.

where carbon dioxide abundance measurements have been taken continuously since 1958. The resulting abundances since that time are shown in Figure 1.1; Figure 11.1 shows the abundance over a longer period. These diagrams, in any of a multitude of forms, have become a kind of centerpiece in the global warming debate and show promise of becoming defining logos of the twenty-first century.

A similar diagram for the Southern Hemisphere comes from the Cape Grim Station on Tasmania and is shown in Figure 11.2. Both diagrams reveal a periodic oscillation with a time frame of one year superimposed upon a rising trend. The oscillations represent the annual growth of vegetation in the respective hemispheres. During the spring and early summer of each hemisphere, vegetation takes into its foliage a considerable amount of carbon from the air and annual decomposition during the fall and winter returns the carbon dioxide to the air.

In latitudes far from the Tropics, the annual peak-to-peak amplitude of the carbon dioxide abundance ranges from 15 to 20 parts per million by volume at high latitudes, down to about 3 parts per million at the equator. This variation reflects the minimal seasonal growth variation in the Tropics. Since about 1960, the amplitude has increased by 20 percent in Hawaii (Mauna Loa) and by 40 percent in Arctic regions. Over the same period, the declining phase (the major intake of carbon by the growing vegetation) has advanced by about seven days. The cause of this finding may be a

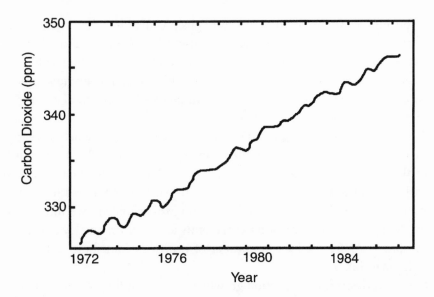

FIGURE 11.2   The atmospheric carbon dioxide over southeastern Australia.

lengthening of the growing season, indicative of overall warming. Here we should remember that a warmer climate and an enrichment of atmospheric carbon dioxide favor plant growth.

The amount of fossil fuels produced every year as oil, coal, and natural gas is well known. Because no major stocks of any of these fuels are being accumulated on a large scale, we can calculate precisely how much carbon dioxide escapes into the atmosphere from burning fuel. To this we may add the product of burning rain forest, although measurements as well as assessment by satellites show that it makes up less than 10 percent of the total. A fair number of almost insignificant minor contributors add to manmade carbon dioxide releases into the atmosphere. However, this is still not yet the total input the atmosphere has to deal with. Two other sources are more difficult to estimate; these are the gas release from the oceans and from volcanic sources. If we leave out the uncertain sources, or at least underestimate them, we arrive at the minimum amount of carbon dioxide passing into the air over a given time interval.

Apart from the yearly oscillations, the measures of this major greenhouse gas in both hemispheres show a steady increase. This must have begun in the last years of the nineteenth century and coincided with the Industrial Revolution. We are reminded of the stories of George Eliot, Charles Dickens, and D. H. Lawrence that describe the hideous downside of burning coal, the scourge of the English Midlands. What interests us here is the present yearly increase, because it is part of the annual manmade input. In fact, because we know the total weight of the Earth's atmosphere, the percentage of carbon dioxide in it, and by how much carbon dioxide increases per year, we can deduce how much new carbon dioxide shows up in the atmosphere every year.

When we compare the annual manmade input with the actual yearly increase in the air, we find the surprising result that only about one third of the input reaches the atmosphere. Where does the rest go and which carbon sinks absorb two-thirds of the annual input? We will look at a number of possible sinks. When analyzing this problem, we must find mechanisms that stow carbon permanently lest it return quickly to the atmosphere. As we shall see, a number of processes contribute to this yet unknown sink.

The yearly cycle of vegetation growth and decay in the Northern Hemisphere is visible in Figure 1.1 of the first chapter. One may also expect to see in this curve the effect of an increase or decrease in the standing vegetation. The total standing vegetation in the Northern Hemisphere was thought to be decreasing as a result of tree diseases, urban growth, and expansion of land for agriculture and urban development. Satellite observations prove otherwise. The reflectance of the Earth's surface in the red and the near infrared is strongly influenced by the green vegetation; indeed, the reflectance in this part of the light spectrum can be used to estimate the total vegetation cover in various parts of the globe. The respective satellite observations, available since 1981 from the satellites NOAA-7, -9, and -11, indicate a substantial increase in the vegetation in the Northern Hemisphere between the latitudes 45 degrees north and 70 degrees north. The longer and hotter summers with their longer growing season, both created by a warmer climate, may account for the phenomenon. But the vegetation increase in the Northern Hemisphere can account for only a small portion of the "missing sink." Even if this sink had taken up part of the human carbon dioxide input into the air over the past few decades, its size is limited: When green covers the Earth's surface, this sink will be full and unable to absorb more carbon dioxide.

Tree trunks may provide a permanent and less limited carbon storage in vegetation. A good part of the wood is carbon, and it will remain so as long as the trunk is alive or safely stowed away as construction material or furniture. When wood decays or burns, its carbon component combines with oxygen and returns to the atmosphere as carbon dioxide.

At present, the biota on land in the Northern Hemisphere remove about 30 percent of the manmade carbon dioxide from the air. Removal by tropical land biota is negligible, as shown through a study of the atmospheric abundance ratio between the elements nitrogen and oxygen (the ratio O_2/N_2). The ratio was analyzed at three sites for 1991–1994. One site is far north at latitude 82.5 degrees north, one is in California, and the third is in Australia's island state of Tasmania in the Southern Hemisphere. The data show that the ratio depends on latitude, varies with an annual cycle, and exhibits interannual variations. Combining the conclusions from this study with those of other carbon sinks requires that an additional sink be found in the Northern Hemisphere: Vegetation cycles that could fill the gap.

The soil, composed of decaying vegetation, is another important carbon reservoir; it contains more carbon than the standing vegetation. The total carbon reserve in soil, worldwide, is difficult to estimate, even more so when we consider its annual or short-term or long-term change. Just think of your own lawn. You know that soil lies just under the grass. You could even measure how many centimeters of soil lie under the grass, and then determine how much of that is carbon. But how much soil and what kind of soil was under your lawn ten years ago, or a hundred years ago? These figures are impossible to come by. And so it is difficult to assess how much of the carbon input into the air has been taken up by a growing soil during the past decades.

In the chapter on ocean currents and their influence on the climate, we will show that all oceans are interconnected by one single conveyor-belt-like system of currents with many up and down streams. It takes a small volume of water perhaps 1,000 to 2,000 years to complete a round trip through the entire system. By that time it has traveled through all the oceans at all possible depths with a temperature range from freezing to tropical heat, and it has also continuously changed its content of organic life.

At this point we should remember that gaseous carbon dioxide tends to penetrate into a water surface and produce carbonic acid. The higher the gas pressure, the more carbon dioxide dissolves in the water. At the same time, the warmer the water, the more difficulty carbon dioxide has in dissolving. When water containing carbonic acid warms up, it expels some of the gas, as you can demonstrate with a bottle of soda.

The uptake of gaseous carbon dioxide from the air naturally happens only at the boundary between water and air, extending this process to a depth of a few meters with the help of the stirring action by waves. The presence of carbon dioxide in the water is a basic requirement for the formation of organic life, just as it is on land. Most marine life forms are restricted to the layers near the surface, and even more so to shallow waters. By a photochemical process, similar to that occurring in plants on land, phytoplankton—the first link in the marine food chain—picks up the carbon dioxide and incorporates the carbon into its structure. We have already seen that two conditions favor the growth of phytoplankton; these are, first, cooler water, which can offer more carbon dioxide, and second, a higher carbon dioxide abundance—hence pressure—in the air. This, however, is not the whole story. To grow, phytoplankton needs minerals as well, particularly dissolved iron. It is well known that some regions in the oceans are rich in nutrients for phytoplankton, yet the minerals phytoplankton require are scarce or absent. The reason for this is the lack of iron.

An experiment was made a few years ago in the eastern Pacific Ocean where seeding of dissolved iron into the water (over an area of about 70 square kilometers, or about 43 square miles) was a striking success. The phytoplankton abundance quickly increased more than twenty times, and the corresponding loss of carbon dioxide in the air above the test area was readily verified. The spectacular plankton blooming that resulted from the seeding suggested a simple way to rapidly remove at least part of the excess carbon dioxide that humans spill into the atmosphere. But the seeding of dissolved iron into the seas can work only up to a point. If nothing eats the excessive plankton, they will die, decay, and convert right back into gaseous carbon dioxide, which then returns to the air. We must ask, too, what other consequences a substantial worldwide increase of the phytoplankton abundance might have. It is too dangerous to count on this method for controlling the buildup of greenhouse gases, even as a temporary measure.

The amount of carbon dioxide that water can dissolve depends on its temperature; this means that all the oceans together can contain a limited amount of carbon dioxide. We are, no doubt, still far from this limit. The oceans, therefore, are the most natural place to look for the sink that must be taking up a large portion of our carbon dioxide input into the atmosphere. We must look for a mechanism of permanent—or of at least long-term—storage of carbon. A number of important points must be taken into account.

First, the interchange between ocean and atmosphere takes place in a thin layer at the water's surface. However, the conveyor-belt system of ocean currents carries the surface water to all parts of the oceans and to all depths while bringing water from the deepest ocean trenches up to the surface.

Second, organic life can lead to a long-term or even permanent storage of carbon. Plankton and other small marine plants and animals contain a considerable amount of this element; but as they die and decay, a process sets in that can be slow enough for bodies to sink or be carried by the currents to depths where the absence of light precludes photochemical processes. Life exists even at the greatest depths, and this form of life must make use of already available organic molecules because they cannot produce any of their own.

More important, however, are mechanisms that lead to a permanent carbon storage. The entire sequence is often termed *sequestration and burial.* Many small marine animals form shells and other solid structures (such as coral reefs) that consist of carbonates and insoluble substances containing a good amount of carbon and that are denser than water. Once its original "owner" dies, the structure, if afloat, sinks to the ocean floor and forms part of the sediment. Entire mountain ranges grow in this way.

Carbonates usually form from remains of once-living animals. However, during particularly high concentrations of carbon dioxide in sea water, carbon can form without the help of living organisms. Because carbonates were particularly abundant at times of very early extinctions, such conditions probably occurred during several periods of glaciation some 600 to 800 million years ago. Therefore, a surge of carbon dioxide from the deep sea, an event similar to the one at Lake Nyos in Nigeria in 1986 (described in Chapter 8), might have been the factor.

We can test in two ways the hypothesis that the overall trend of rising atmospheric carbon dioxide abundance is linked to human input. The amount of carbon from fossil fuels being burned each year is well known. To this we should add a small quantity to allow for the burning of vegetation, mostly in the tropical rain forests, plus a natural component from volcanic sources. The total amount can then be compared to the additional carbon dioxide appearing in the air.

A second check looks at the abundance of the carbon isotopes C_{13} and C_{14}, which are formed in the air by interaction with cosmic radiation; from these abundances, we can calculate the portion of this gas that comes from natural causes. These isotopes have a half-life of about 6,000 years: Every 6,000 years, half of them return to their normal atomic form of carbon, the isotope known as C_{12}. This fairly long half-life means that the relative abundances of carbon isotopes found in plants are nearly the same as those found in the atmosphere. However, because the carbon contained in coal, oil, and natural gas lies deep under ground, well protected from cosmic radiation over millions of years, it has lost almost all of its unstable isotopes, C_{13} and C_{14}. For this reason, the injection of carbon into the air from fossil fuels tends to lower its unstable isotope abundance. Such studies have been carried out for a number of years at several sites around the world. They serve not only to check on the origin of the atmospheric carbon but to permit the study of the mechanisms that carry the carbon dioxide from the emission centers to other parts of the world, in particular from the Northern Hemisphere, where most of it originates, to the Southern Hemisphere.

Carbon dioxide penetrates sea water only to a depth of a few meters, where much of it can be absorbed by plankton. The plankton, which form the bottom of the food chain, if not consumed by higher forms of life eventually die and sink to the ocean floor, taking their carbon with them. Some of this carbonate may be turned back into carbon dioxide through decomposition. Ocean currents carry carbon from one part of the sea to another and eventually down to all depths, where it must be sought to complete the carbon picture. No complete tally is available yet, but a fleet of several ships is presently scanning the waters of the North Atlantic to check the carbon balance throughout the year. Similar studies are needed in other oceans before the role of ocean currents in the carbon cycle can be fully understood.

As we mentioned earlier in the chapter, water has a limited capacity to absorb carbon dioxide—the capacity diminishes with rising temperature—even if we take into account extraction by organic matter. If the greenhouse effect causes air temperature to increase, sooner or later the temperatures of the oceans will also increase. When this happens, and if carbon dioxide already saturates the surface waters of the oceans, they will begin to release some of it into the atmosphere. Even if this does not happen, the oceans will eventually be unable to store excess carbon dioxide, with the same outcome. Because we do not know enough about the content and distribution of carbon dioxide in the oceans to predict its future, it is not reasonable to simply extrapolate the curves in Figures 1.1 and 11.1 into the future in order to calculate the contribution of carbon dioxide to the global warming in the years to come.

The contribution of a particular gas to the absorption of infrared radiation depends not only on its abundance in the air but also on the ability of its molecules to transform infrared light into some other form of energy. Carbon dioxide is still the most important greenhouse gas, but only because it is far more abundant than any of the others. Its molecules are not particularly efficient at absorbing infrared radiation, and are certainly less efficient in this role than are other trace gases also present in the air. Carbon dioxide is also nearing its maximum absorption capacity.

Methane (CH_4) is the second most important infrared absorbing gas and is at present the gas with the most rapidly increasing abundance, at a rate of nearly 1 percent per year. Human activity does not directly cause the methane increase, but indirectly we are partly responsible. At the moment, the main producers of methane are rice paddies and cattle flatus. Leakages from distribution systems of natural gas and garbage dumps also spew methane to the atmosphere. These sources are related to human activity, and some of them are even now being tapped for energy production. Farmers commonly extract the gas from cow manure for fuel, and some cities extract methane from their garbage dumps for use in industry. Swamps are among the natural sources of methane, hence its alternate name of marsh gas. The contribution from termites is not yet important; however, termites are among the world's fastest growing animal species, and their contribution may yet become significant.

The nitric oxides are among the next most important greenhouse gases. Combustion at high temperature in the engines of automobiles and aircraft can burn some of the nitrogen in the air, thus producing nitric oxides. These gases are also released by some of the combustion processes and chemical reactions used in industry. Scientists estimate that some 40 percent to 70 percent of the nitrogen oxides emitted into the air are of human origin. These oxides, as long as they remain in the gaseous phase, are efficient infrared absorbers.

The chlorofluorocarbons (CFCs), also known as *freon gases,* are notorious for their destructive effect on the stratospheric ozone layer because they are among the most effective infrared radiation absorbers. One CFC molecule absorbs as much as 30,000 carbon dioxide molecules. This, together with their long lifetime in the troposphere, ranks them among the significant greenhouse gases in spite of their relative scarcity. They are used as propellants in sprays, as cooling agents in air conditioners, and for the expansion of foam. By the late 1980s, the yearly production had reached levels of more than half a million tons per year worldwide, but fortunately they are now being phased out in most countries. Since 1995, their use has been restricted to medical applications. Their presence in the atmosphere will be felt for decades to come, after which time the ozone layer will begin to restore itself.

Ozone, both in the stratosphere and in the troposphere, makes important contributions to the atmospheric radiation balance because it absorbs infrared as well as ultraviolet radiation. At the same time, unlike the other greenhouse gases, ozone is a toxic gas. We will discuss ozone further in a later chapter. We need only point out here that the gas is now diminishing in the stratosphere, where it is needed, and simultaneously increasing in the troposphere where, unfortunately, it is not. Human activity is causing both changes.

All the greenhouse gases have one feature in common: Their molecules consist of three or more atoms. The most pervasive of them all, water vapor, has not been mentioned in this context. The abundance of water vapor in the air is, as already explained, a function of the temperature and pressure and the abundance of condensation nuclei, along with the sources of condensation. Water vapor can rapidly change from its gaseous phase to the liquid or solid state; its abundance in the air and its contribution to the greenhouse effect is therefore highly variable.

Water vapor also causes a distinct action, called *positive feedback*. An increase in atmospheric temperature caused by the greenhouse effect tends to lower condensation and increase evaporation, thus adding more vapor to the greenhouse gases and enhancing their effectiveness. Locally, the amount of water vapor in the air can be measured with psychrometers. Radiometers measure the amount of solar radiation lost in the atmosphere in the spectral regions where water vapor absorption is most effective. These measures permit the calculation of the total amount of vapor only along the path of sunlight through the atmosphere. Even so, such measurements are of local significance. Worldwide data are now being collected from satellites, but practically no global water vapor abundance data are available for the pre-satellite era.

The present greenhouse effect is the result of the combined influence of all the gases in the Earth's atmosphere whose molecules are triatomic or larger. The diatomic molecules such as nitrogen and normal oxygen do not make a contribution. If there were no greenhouse gases at all, the global temperature would be at least 10° to 30°C (18°F to 54°F) cooler than it is at present. The carbon dioxide contribution alone still accounts for over half the sum. Methane and the other gases, however, are expected to increase rapidly and will soon form the majority of the atmospheric effect on global warming.

Of the engines that drive the climate variations on the Earth, one could obviously be the Sun. Measurements of the total solar radiation flux from satellites have been possible only during the last few decades. These measurements show changes in the solar irradiance, but mostly of marginal significance, less than 1 percent of the total. The time span covered by observations is too short for us to reach conclusions about periodic or nonperiodic variations in a time scale above a decade. Here, we must seek variations with time scales from centuries to millennia and beyond.

We can compare observations of solar irradiance with the respective meteorological data. Temperature variations predicted from the calculated and observed solar irradiance are shown in Figure 11.3. There appears to be a sign of solar radiation in the terrestrial climate record but no firm conclusion is yet possible.

It is more difficult to detect long-term variations in the solar irradiance. Pre-satellite data are too uncertain for two reasons. First, we cannot calculate precisely the atmospheric extinction, the loss of flux in the atmosphere caused by absorption and scattering of the incoming solar radiation. The extinction depends on the wavelength, the position of the Sun in the sky, and the composition of the air at the time of the observation, including its highly variable water vapor content. Large-scale volcanic eruptions such as that of Mount Pinatubo in June 1991 can cause a sudden and unexpected input of gases and particles into the stratosphere; such an input can increase the extinction of sunlight through the atmosphere. At the zenith, the extinction may exceed 20 percent of incoming light, and the losses greatly increase at higher zenith angles. It usually takes a long time, too, before this volcanic material is more or less uniformly distributed throughout the atmosphere, and it takes years before condensation processes remove it completely.

Calibrating the receivers used for measuring the solar radiation for sensitivity presents an even more difficult problem. The calibration can be accomplished only through a comparison with a standard light source; the constancy of the source must be maintained over time, which adds to the difficulties because no such source is available at this time. We have already mentioned that a variation of the solar irradiance with the eleven-year sunspot cycle is known. Accurate records of the number and size of sunspots have existed for more than a century, and cruder information for

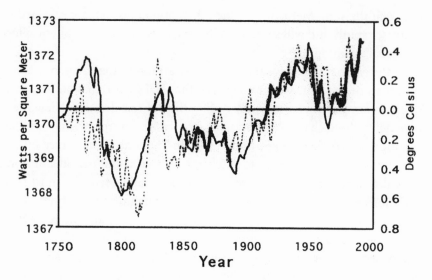

FIGURE 11.3   Combined solar irradiance model and Earth surface temperature variations.

several centuries. Sunspot activity varies from cycle to cycle, and long-term variations have been found. If, however, the detected correlation between sunspot numbers and irradiance holds at all times, the solar-driven temperature variations are too small to explain global climate variations. Any other solar irradiance variation not reflected by the spot activity, naturally, cannot be excluded. Only satellite data over several more decades will be able to show whether such variations exist.

Sunspot activity and other features related to it (the solar magnetic field, for example) are not the only measurable parameters that may prove to be reliable indicators of the solar irradiance. Another possible measurable quantity may be the solar diameter. For about two centuries, scientists have tried to detect possible variations in the Sun by attempting to measure its diameter with accuracy. Direct measurements of the diameter with astronomical telescopes, calculations using the transit time of the planet Mercury to pass in front of the solar disk, and the duration of solar eclipses have all been used for this purpose with no success, the main reason being the difficulty of defining the edge of the Sun.

A new instrument, similar to a sextant, can directly measure the diameter of the Sun at any angle. So far, the instrument has been used on unmanned balloons so they can avoid at least most of the effects of turbulence in our atmosphere. Eventually, the instrument will be flown above the atmosphere by rockets or satellites. Several more decades will pass before we can tell with certainty whether solar diameter measurements are a reliable indicator of the solar irradiance. If this turns out to be so, then diameter measurements of the past, naturally within their uncertainties, can be used to reconstruct the history of the solar irradiance perhaps for as much as a couple of centuries. Variable irradiance is such a common feature among stars that it would be no surprise if the Sun was found to be slightly variable as well. By how much and with what period it might vary are questions that observations alone can answer.

Cyclic features in the global meteorological system of the Earth have long been sought. In particular, weather records have been examined to try to detect variation that could be linked to the sunspot cycle. As we have already mentioned, it appears that more than just the solar cycle is involved. Only in the stratosphere has a clear relationship with the solar cycle been detected, but how this relation works its way down into the lower atmosphere is not yet clear. Two other cyclic effects have been found in recent years: One is the so-called El Niño Southern Oscillation (ENSO) and the other is the Quasi Biennial Oscillation (QBO). We will discuss these in a later chapter.

Worldwide meteorological data with sufficient coverage in time and space are available only for the last few decades; it is therefore obviously difficult to firmly establish the presence of any of these periods—the solar cycle, ENSO, and QBO—in any of the meteorological parameters such as global or local temperature, precipitation, etc. It has been firmly established, though, that major volcanic eruptions have a noticeable effect on the meteorological conditions in many parts of the world. In the previous chapter, we mentioned an effect that can be attributed to the eruption of Mt. Tambora in 1815. The eruptions of Mt. Agung in 1963, of El Chichon in 1982, and of Pinatubo in 1991 were all major events, though not quite as large as that of 1815. It will take several more decades of observations undisturbed by volcanic events before the periodic events can safely be separated and understood.

The injection of aerosols, either of volcanic or human origin, is believed to produce a global cooling of the planet, but the injection of greenhouse gases is expected to produce a global warming. The cooling effect by the aerosols works through two mechanisms. To start with, particles scatter sunlight, returning part of it back into space. Water vapor can condense on the particles, making even larger particles; these contribute at the same time to cloud formation and to the extraction of water vapor from the air, thus reducing this component of the greenhouse gases. The reflecting properties of clouds are also altered by aerosols. Major volcanic eruptions eject particles and gases into the stratosphere, where their presence may cause colorful sunsets. It was once believed that most of the particles ejected by volcanoes into the stratosphere consisted of ashes and dust. Following the eruptions of El Chichon and Pinatubo, however, it was found that most of the stratospheric particles are sulfuric acid; the sulfuric acid is created from the emission of sulfur dioxide, which then combines with atmospheric water vapor to form the acid. Satellites are now documenting the distribution of particles and gases in the air. Figure 11.4 shows the global distribution of ejection material from Pinatubo, taken five months after the eruption. The vertical distribution of these particles can be determined directly from the ground by the LIDAR-stations, of which there are several around the world. Laser light pulses emitted from the ground straight up into the air are reflected and scattered by the particles. The time the echo requires to reach the emitter contains information on the particles' elevation above ground as well as on their distribution in depth. An example from Canada, again showing the Pinatubo ejecta, is shown in Figure 11.4; these particles are expected to circle the Earth for at least two more years; at the date of the eruption, particles from the El Chichon eruption had just barely faded away. Theories indicate that they should cause a global cooling of about half a degree. Eventually, these particles will sink into the troposphere, where condensation will bring them to the ground.

The curve appearing in Figure 10.2 is about the best representation of global temperature during the past century that can be produced at present. Although it exhibits semiperiodic and erratic variations superposed on what appears to be a rising trend in accordance with what is to be expected, researchers cannot yet determine whether the rising trend is real. It could well be caused by the accidental accumulation of various periodic effects, with erratic contributions from volcanic eruptions and other sources. It has taken a long time to determine how the global climate reacts to human activity. The disturbance caused by Pinatubo disappeared only

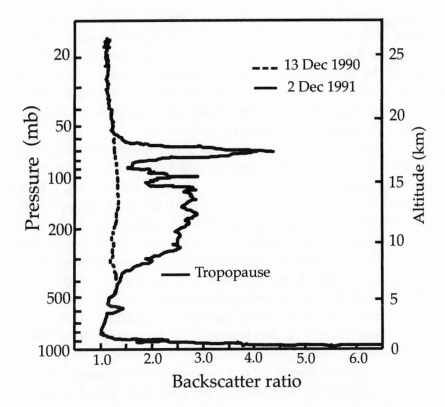

FIGURE 11.4   Distrubution over northern Canada of Pinatubo ejecta by altitude.

very recently. Only now, therefore, are we able to look through an atmosphere undisturbed by volcanoes. If no new volcanic eruptions occur, we will be fortunate to enter a period free of their influences.

The oxides of sulfates and nitrogen, once injected into the air, readily combine with water vapor to form the nitric and sulfuric acids that then become condensation nuclei, thus producing acid rain. Note that the acidity of the rain drops is so low that they do not constitute a serious problem. Wherever the water from acid rain evaporates, however, it leaves the acid behind. It was found that an acid raindrop on a leaf may be converted into a highly concentrated acid, albeit of minute dimensions, when the water has evaporated; the concentrated acid may then burn a hole into the structure of the leaf. It becomes evident that acid rain accumulates acids in the soil and in lakes and rivers. Water's acidity or alkalinity can be measured and expressed in numerical terms on the pH scale, defined such that a value of 7.0 is assigned to a neutral solution, values less than 7.0 mean acidity, and those in excess of 7.0, alkalinity. Normal pure rain water is expected to be slightly acid because it will take up some of the atmospheric carbon dioxide and convert it into carbonic acid. In the Tropics, where there is still little effect of worldwide contamination by acid substances, the pH values normally scatter around 6.5.

The acidification of the soil, lakes, and rivers brought about strict measures to control the emissions that cause acid rain. Catalyzators for car engines and industrial smoke stacks were invented, and their use is being enforced in many parts of the world; unfortunately, many third world countries have not yet taken steps to control exhaust emissions. The effect of these measures is noticeable in places that were once known for their extremely high air pollution; for example, London and Pittsburgh. Other large cities, for instance, Mexico City or Santiago, Chile, have so far been unable to solve their ever growing pollution problems.

The control and elimination of manmade air pollution can also have negative consequences. As already mentioned, the presence of aerosols in the air brings about a cooling effect. Possibly worldwide air pollution caused by particles has so far compensated for greenhouse warming; therefore, cleaning the air could give the greenhouse warming a better chance to work.

Much has been said and written about the effects of burning tropical rain forests. To understand the situation, we have to make a rough balance of carbon reserves and their distribution over the Earth. We shall leave aside the carbonates because they are in a type of permanent storage from which they cannot easily be extracted. If we take the carbon contained in the atmosphere as carbon dioxide and methane—carbon monoxide (CO) is negligible for this purpose—as the unit, we find that the standing vegetation contains a little less than one unit and vegetation decomposing in the soil contains about two units. The oceans, counting all depths, store about fifty units. The fossil fuels still underground in the form of coal, oil, and natural gas represent about ten units. Carbon is constantly being cycled from the air into the vegetation, and from there into the soil, where it remains somewhat longer, and then back into the air. When we burn vegetation, we simply shortcut the soil phase; that is, carbon travels from standing vegetation right back into the air. If the burned land soon recovers its vegetation, it takes the carbon out of the air again, and in the long run nothing has happened. Things are different when the ground remains cleared; that land cannot take carbon dioxide back out of the air and what was already in the soil as decomposing vegetation could eventually find its way back into the air. The second scenario happens in most cases.

The amount of carbon dioxide released into the air as a result of burning the rain forests is small when compared with the amount released through the burning of fossil fuels. Fossil fuels make up at present about 7 percent of the total carbon dioxide input into the air, and only a small part of that total comes from burning the rain forests. In many parts of the world, wood is still a primary energy source for households; the carbon dioxide input from wood, however, is not the major concern. Vegetation (and in particular trees) determines the evaporation rate of water in a forest. Trees not only protect the soil from direct sunlight, they also store large amounts of water; they release water as needed as a means of controlling their own temperatures. Soil temperature, and therefore air temperature, humidity, and precipitation, is greatly altered when a forest is destroyed and the land is cultivated or left bare. Such alterations affect not only the local climate: Winds carry the effect to other areas as well.

Rain forests are burned for three reasons. One is to create new areas for agricultural purposes. Rain forests usually have only a few centimeters of soil because everything that falls to the ground and decomposes is quickly recycled. When the soil is exposed, the minerals present in the ashes make it fertile for a while, but after a couple of crops, or at the most three crops, the soil is exhausted. Even fertilizers don't help much because the soil soon falls victim to erosion from water and wind. What is left behind, then, is sterile land that won't recover for hundreds or thousands of years, if ever. The farmer simply moves on and clears another piece of virgin forest. The process has affected large areas of the Amazon River basin, and every year the rate of destruction increases. This method of agriculture is the only way people native to these areas can subsist; the situation will not change until human society offers them an acceptable alternative.

A second reason is gold mining, which destroys forests in northern Brazil and southern Venezuela. When miners find gold in the sand beneath the soil, they not only burn the vegetation but remove the soil and divert rivers to wash out the gold. To precipitate the gold, they add mercury to the water. The mercury not only poisons the flora and fauna downstream, when it reaches the oceans, it continues its rampage. Back at the mining sites, miners leave behind a scene of total destruction.

A third reason is the "need" for tropical wood, particularly in the Far East. Large areas of forest must be cleared to reach and haul out these lone-standing giant trees; they wind up as fancy pieces of furniture in the homes of well-to-do people in far away countries.

It may seem strange that the real extent of the destruction remained unknown until satellites could deliver pictures that covered the entire globe. The twenty years or so of satellite control are already enough to show the dramatic and rapid recession of the rain forests: Today, not much is left of areas once covered by dense woods. Estimates show that all rain forests will disappear within less than a century if the present rate of destruction continues. We have already pointed out that the social and economic structure of third world countries provokes this destruction. Only a common effort by all mankind can save the forests.

Deforestation is not new. The woods of Greece and of many of the Mediterranean islands were destroyed in ancient times by people collecting firewood and by the overgrazing of domestic animals, notably goats. The loss of the forests in northern Chile and in Peru is more recent. The Spanish conquerors found forests all along the South American coast from Lima to Valparaiso, and dense woods in the interior. Here, in the forests the main reason for cutting down the trees was the production of charcoal for the silver mines in Peru. Remains of tree stumps can still be found in the Atacama Desert, now void of vegetation. A situation has developed along the border between the Sahara Desert and the Sahel just south of it. From their former way of life, the inhabitants of the area switched to raising cattle on a scale that cannot be supported by the vegetation. As a consequence of the overgrazing, the soil has dried out and scarce trees have disappeared. As the land turned into desert, the inhabitants moved south, where they found still better grazing and so repeated the damage. During the last four decades, the desert has moved southward at a rate of up to 10 kilometers (6 miles) per year. The altered surface conditions have reduced the formation of precipitation because rivers and lakes began to dry out. The recession of Lake Chad in Nigeria began around 1960; if the lake continues to recede at its present rate, within a few years it will disappear altogether.

To all this we have to add the consequences of population growth the world over. New homes must be built, along with new roads and highways, new public buildings, airports, etc. It is estimated that every year an area about the size of Switzerland is taken from the wild to make room for construction. Garbage dumps, including those that store atomic waste, are expanding at an impressive rate. Human contamination of air and soil takes its share of the wild. In many parts of Europe, more than 50 percent of the standing trees are damaged or dead. The agony of trees has for many years been attributed to the acid in rain, which works its way into vegetation either through the roots, that is, through the soil, or through the leaves. It now appears that tropospheric ozone, also a manmade and undesirable product, plays a major role in damaging trees. The net effect is again receding forests. In Europe and in North America, however, forests are recovering at high latitudes, where vegetation has increased over the past years.

We might expect an enrichment of atmospheric carbon dioxide and a warmer climate to favor vegetation, especially if we take into account the increases in global evaporation, and therefore precipitation, that a higher global temperature would cause. Experiments to determine how growth is affected by carbon dioxide concentration, all other parameters being equal, show that during recent years the total biomass in Europe in the form of standing vegetation, that is, trees, has increased. In fact, some investigators suspect that the missing sink of carbon dioxide might be the vegetation. Whether pollution-control measures, which came into effect recently in Europe and in North America, or more abundant carbon dioxide have helped the biomass is still an unanswered question. If trees are supposed not only to take up the carbon dioxide not accounted for but also to make up for losses of wilderness, recent tree rings should show a noticeably increased width; but this does not appear to be the case.

The behavior of ice sheets in the Arctic and Antarctic is an important aspect of climate change. As we have already mentioned, from a theoretical point of view it is not yet possible to conclude how the ice sheets respond to global warming. As there appears to be a rising trend in the global temperature, at least at present (as shown in Figure 10.3), can its effects already be observed at the poles? Only satellite photographs can tell the complete story, and naturally they do not cover a long period. The present data indicate a significant reduction of the floating ice around the North Pole, yet at the same time Greenland's ice is growing. Although more snow appears to accumulate in the Antarctic, observations show a breakup of the floating ice shelves. Only the years to come will tell us whether these phenomena are a long-term feature.

How does the temperature of the ocean surfaces respond to the present global warming trend of the atmosphere? Data taken from ships are too scarce, too far spread in time and space, and often not accurate enough to give the answer. Satellites provide data that are global and far more accurate, but only since about 1990. An upward trend of about 0.1 degree per year has been found, although some authors challenge this result. Only future data can show whether this trend is real and whether it will persist. Warming of the waters should lead to an expansion of the liquid body of the Earth, and therefore to an increase in its diameter. We must recall here that the temperatures measured refer only to the surface. A permanent sinking and upwelling of water masses occurs through the ocean currents, and these will eventually transmit the temperature increase at the surface to all levels of the oceans. Naturally, this will take a long time.

We should mention here that because a heat source at the bottom of an ocean cannot be excluded, the measured surface temperature increase does not necessarily reflect a temperature variation of the entire water body; indeed, the variation could be larger or smaller than that presently observed at the surface. Diameter measurements of the liquid Earth can now be made from satellites with great accuracy. Such data are available for several years; a growing rate of a few tenths of a millimeter per year is the surprising result. No conclusion can be reached as long as no complete three-dimensional temperature data are available, one of the major objectives of present ocean research. Even so, it is clear that the diameter measurements contribute another climate-related argument.

We have stated in a previous chapter that the effect of urbanization can make the information of a long-standing meteorological station next to useless in forecasting long-term climate changes. We might even suspect that all large cities seen as a whole might have a noticeable effect on the global climate. But studies show that the effect of all cities combined is still negligible when compared with the other mechanisms that possibly or certainly have a detectable influence.

Ancient and not-so-ancient meteorology was based mostly on empirical relations without any understanding of the underlying causes. Modern meteorology still makes use of such empirical relations. In times past, it was always thought that weather conditions in two regions distant from each other were unrelated except, naturally, for the common seasonal variations and those caused by a global intrusion into the system (such as large-scale volcanic eruptions). A low- or high-pressure system dissipates long before it has a chance to circle the globe, and propagation across the tropical regions into the other hemisphere seemed even more unlikely. When weather balloons began to climb into the stratosphere, it was found that even that far up there is "weather" that changes from day to day and across the year, and that it does not necessarily reflect what is going on in the troposphere. Yet an interrelation was soon found. The so-called jet streams—stratospheric high-velocity winds with relatively small extension in height and width—were discovered and analyzed. (The jet stream can even be observed through astronomical telescopes.) The location of the jet streams—one is present in each hemisphere at relatively low latitude, and often each splits into two separate streams—determines the course of tropospheric weather systems. It is quite common now to show the location of the jet streams on weather maps. The worldwide effect of El Niño or of a QBO may also be transmitted by the stratosphere over large distances. We discuss these climatic features in Chapter 13.

To a long list of correlations between meteorological phenomena from old times, many new ones have been added, and others are being discovered all the time. They constitute a challenge for theoretical meteorologists, who have to find an explanation for them. Among the newly discovered relations, we find one that relates the weather conditions in West Africa to the area where hurricanes strike the United States. There are dry and wet years in West Africa. When West Africa is wet, more hurricanes develop in the Caribbean, the Gulf of Mexico, and the tropical Atlantic; the majority of them move along the eastern seaboard of the United States, with only a few striking the Gulf Coast. When West Africa is dry, fewer hurricanes develop and just as many move into the Gulf region as they do along the Atlantic Coast. This relation, as mysterious as it may be and influenced somehow by El Niño, is obviously a useful tool for hurricane forecasting. During an El Niño period, more hurricanes develop over the Pacific and fewer over the Atlantic. Another relation, probably more readily understood, is that the temperature of the South Atlantic determines precipitation in the Sahel. A relatively warm ocean means drought for the Sahel; a cool ocean brings plenty of rain into the area, again a feature useful for prediction.

Theoretical calculations show that expected temperature changes caused by the greenhouse effect will be the greatest near the poles and minimal in tropical regions. Observations made at intermediate and high geographic latitudes, therefore, receive particular attention. So far, the effect of global temperature rise, if present at all, is too small to be safely separated from the large fluctuations, commonly referred to as *noise*. By noise, we simply mean the large day-to-day temperature variations that changing weather conditions create. In the Tropics, noise is considerably less because virtually no seasonal temperature variations occur, and day-to-day variations are small. Data from equatorial stations may be more suitable for the determination of a global temperature trend than those from other latitudes.

12

OCEANS:
THE MAJORITY RULES

Beyond all things is the ocean.

SENECA

OCEANS COVER OVER 70 PERCENT of the surface of the globe. If we add water in solid form in the ice caps of Greenland and Antarctica and on top of the seas nearby, the total is somewhat higher. In classical times, ocean was thought to surround the land on all sides. For those regions explored and known at that time, the majority of the continents of Eurasia and Africa, this was true, and it is true today. Ocean was frequently worshipped as a god by past peoples, and it is not surprising that Sigmund Freud and others described a person's contented feeling of oneness with all creation as "oceanic."

Water has a most unusual property, unlike that of almost any other substance. Most stuff continues to contract with colder temperatures; the greater the chill, the smaller the volume occupied, and thus the higher density of the coldest part makes that part sink to the lowest possible level. Water, however, reaches its greatest density at a temperature of 4°C (39°F). Then with increasingly colder temperatures, it expands; at the freezing point, it expands at a considerable rate. We know from this that the temperature at the lowest point in a large body of water is near 4°C; at this temperature, water is at its densest.

We can be thankful that this is so. If water behaved the way most other substances behave, ice would form first at the bottom of a body of water, not at the top. Skaters would have to find or manufacture a rink or pond frozen from top to bottom. Ice cubes in drinks would sink to the bottom of the glass. And perhaps the Titanic would still be sailing today, because icebergs would form at the bottom of the sea, not near its surface.

But the expanding ice feature of water is far more important even than this, for it might not have provided the benevolent environment early life needed to form and reproduce. As the air temperature fell below freezing, the top layer of the pond would have frozen and sunk, allowing the next layer to do the same. As the pond froze solid, life in it would not have made it through the winter. As the pond warmed in the spring, a layer of water would have formed on top of the ice and would then have vaporized into the air. As the water thawed, it would have continued to vaporize until the pond disappeared. Such conditions would not harbor life for long.

The climate at any point on the Earth's surface is to a first approximation determined by the total amount of solar radiation received in a day during a certain season of the year. The variation of this total over a year then makes up the main ingredient of the climate at that particular site. The total radiation received during an entire year strongly depends on geographic latitude. The geometry involved here is easily understood, and we shall not go into it further at this point. Latitude, however, is only part of the story, as we know from the comparison of the climate of western Europe with that of the northern United States and southern Canada at corresponding latitudes. We all know that the principal reason for this difference is the Gulf Stream, which carries warm water from the Gulf of Mexico across the North Atlantic Ocean, bathing the entire west coast of Europe from Portugal to Iceland and Norway in relatively balmy water. Eastern Canada and the northeastern United States, however, experience the cold Labrador Current, and Labrador, lying over the same latitude range as do the British Isles from Land's End to the Orkney Islands, supports only about 30,000 people, as opposed to Britain's 60 million people living in an area of about the same size. This demonstrates the decisive effect of ocean currents on continental climates.

Even more striking proof can be found concerning the importance of the oceans in the global climate system. Were the climate dictated exclusively or even mostly by the amount of solar radiation received, we could calculate the average yearly temperature expected at any given latitude. Above all, we would know what temperature difference to expect between the equator and the poles. The observed temperature difference is considerably less than this expected temperature. Because the temperature at the poles is much higher than would be expected from incoming solar radiation only, there must be an effective energy transport from the equator to the poles. In principle, the atmosphere or the ocean currents or both can perform this energy exchange. The Gulf Stream is just one of the means by which energy is moved from the Tropics to the Arctic and Antarctic. The heat of the Gulf of Mexico is transferred to the air all along the way, but mostly where the current meets the frigid arctic air; by that time, the Gulf Stream has become a current of very cold water. This all looks very simple and convincing, yet it leaves a number of important questions unanswered. First, where does the water carried by the Gulf Stream to the Arctic Ocean come from, and where does it then go? Second, did the Gulf Stream always exist, and will it continue to exist? If not, could a failure and subsequent reappearance of the Gulf Stream have something to do with the ice ages?

It is obvious that fresh water must flow into the Gulf of Mexico from somewhere. Even the Mississippi and other large rivers provide nowhere near enough water to replace that lost by the Gulf Stream. At the same time, water must flow out of the Arctic Ocean to make room for the water being brought in by the Gulf Stream, and the same condition must apply to all of the other ocean currents (the major ones are located and named in Figure 12.1). Thus an ocean current has to be part of a worldwide system of currents working together like a conveyor belt.

Theoretically, each of the Earth's oceans could have its own conveyor-belt system of currents; but as it turns out, all the currents are interconnected by one single system, shown in Figure 12.2. The currents are by no means all surface-water currents. Some are often of surprisingly small vertical extent at all depths and cross each other at different levels. These vertical cross-currents can be so strong that they have been known to shear cables carrying equipment to the depths. The amount of water the currents carry, the flow, and the speed at which they move far exceed the flow of the Amazon River. The Amazon is by far the largest source of fresh water flowing into the oceans, and by itself accounts for one fifth the outflow of the world's rivers.

To halt the system, or even a part of it, is impossible because it would involve every current in every ocean. This means that the Gulf Stream has endured for many millions of years, and that it will endure for many more. That a current could shift into an adjacent area seems much more feasible. Evidence exists in the North Atlantic Ocean for such shifts in the past, and they may be related to the glacial periods. We could speculate that a shift in the Gulf Stream to the south or downward into the deep sea at a lower latitude might be enough to turn on the freezer for European climates; such a

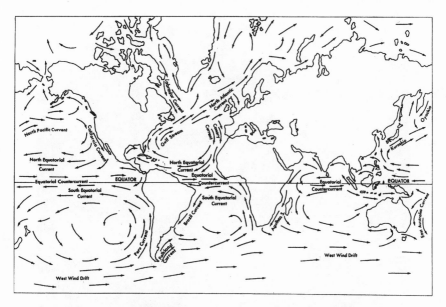

FIGURE 12.1   Ocean currents showing the "conveyor-belt model."

shift could be a possible origin of the Little Ice Age of the past several centuries. A shift in the Gulf Stream would hardly affect the climate of North America, however, and even less that of the Southern Hemisphere. That all glacial periods or ice ages occurred simultaneously throughout the world shows that their origin is not caused by the shift of a specific ocean current; there has to be a more basic and profound cause for the climatic changes the Earth has undergone.

What keeps the ocean currents going in their present places and at their present speeds? There should be enough friction with the ground, with the coastlines, and with stagnant waters to bring the entire system to a halt unless other forces keep them moving. Because the currents don't constitute a perpetual-motion machine, they must respond to an energy source—but which one?

The wind is the most obvious motor driving ocean waters, though by no means the most important. Wind affects only the surface waters, and a prevailing wind direction gives rise to a permanent motion of the surface waters in the same direction. Strong winds can push an enormous amount of water; for example, a typical gale over the North Sea with westerly winds piles water against the coasts of Belgium, the Netherlands, Germany, and Denmark. On such occasions, high tides may rise up to ten me-

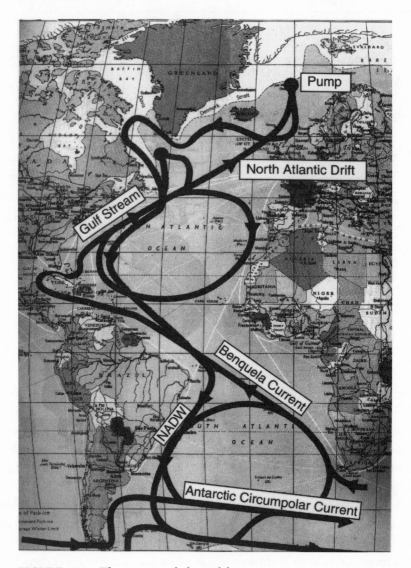

FIGURE 12.2   The conveyor-belt model.

ters (32 feet) above the normal level and flood ports as far inland as Hamburg, 60 miles (about 100 kilometers) from the mouth of the Elbe River. (This has happened four times since 1950.) The Gulf Stream is also helped along by the generally south-westerly winds over the midlatitudes of the North Atlantic Ocean.

Before the satellite era, wind data over the oceans came only from on-board observations on ships and from weather stations on islands and along coasts. By measuring the scattering of radar signals caused by the wavy structure and motion of the sea's surface, satellites can now give a permanent and detailed picture of the wind pattern over the entire liquid surface of the Earth. Japan's Advanced Earth Observing Satellite can, for example, produce a detailed wind picture of all the oceans, extremely useful for the detection and surveillance of typhoons and hurricanes.

Of greater importance are the mechanisms that in the first instance produce a rising or sinking motion of water masses. Light water, water of less density, floats on top of denser and therefore heavier water. If the water on the surface becomes denser, it sinks to the bottom and is replaced by the now lighter water underneath. Several mechanisms can change the density of ocean water.

Two parameters, temperature and salinity, determine water's density. The warmer the water, the lower its density; this helps the warm currents in the Tropics maintain themselves on the surface. As water cools, it becomes denser; the maximum density is reached at a temperature of 4°C (39°F). When a surface current reaches the polar regions and cools to near 4°C, it sinks to make room for water of lower density. When cold water rises to the surface and warms, it expands and becomes lighter, thus tending to stay at the surface; there it warms further and becomes even lighter.

A change in salinity is parallel to the change of density caused by a temperature change. The greater the salinity of the water, the higher its density. Warm water under a dry atmosphere can evaporate efficiently. Evaporation leaves the salt behind in the water, which then becomes saltier and therefore heavier. The rate of evaporation, and with it the rate of increase in salinity, depends on the temperature of the water and on the humidity of the air above it. When ice forms, the salinity also increases because most of the salt remains in the liquid water. The influx of fresh water free of salt, from rain or snow or rivers, tends to reduce the salinity of ocean water and thus reduce its density.

The complex of mechanisms that sets the ocean water into motion is called the *thermohaline circulation,* and we are at present considering just one operation in the entire circulation system. Although the present global climate is decisively influenced by this circulation system, evidence shows that in other times the global climate system was different, at least in the details. The overall system, however, has changed little, if at all.

We can now ride along on the Gulf Stream from its origin in the Gulf of Mexico to its end in the Arctic Sea. It starts as a stream of warm and light water. As it crosses the ocean, it continuously yields energy to the atmosphere and consequently cools. All along the way, water vapor evaporates into the air and increases the density of the liquid water; at the same time, incursions of fresh water from rain or rivers along the European coast counter the effect. Once the Gulf Current has reached polar latitudes, it has given up all its extra energy and is so heavy that it sinks into the depths of the ocean. The present models suggest the action of a gigantic oceanic pump as it sucks water to the deeper layers; the sucking action can be likened to that within the drain of a bathtub.

Figures 12.1 and 12.2 reveal the present status of the currents in the Atlantic Ocean, including the location of two "pumping stations." The global panorama of ocean currents is also shown. A displacement to the west of the North Atlantic Drift or a failure of the pump in the Greenland Sea would keep the warm current away from Europe, resulting in a drastic change of the European climate. This type of event is suspected as the cause of the Little Ice Age. Sometime around 1400, the Gulf Stream moved southward and aimed toward the Bay of Biscay, just to the west of France. The British Isles, and even more Scandinavia, Iceland, and Greenland, were denied the benevolent warmth that embraced them before that time. If this was a gradual process, Greenland would have been affected first, as we now know to be the case. After some four centuries of the colder climate and for reasons unknown, the Gulf Stream reverted to its former path.

Wind and density changes are not all that keeps the ocean currents going. Just as it does in the atmosphere, the Coriolis Effect produces rotating gyres in the oceans. These gyres are permanent features in the ocean-current system, one being located in the North Atlantic Ocean.

In summary, ocean currents are driven by density changes in the water. Evaporation causes the salinity and density to rise; precipitation and the insurgence of river water may reduce them concurrently. These processes may be going on simultaneously at one place or another, causing water masses to sink or rise. Wind then sets them into motion, the Coriolis force changes their course, and the combined effects produce something akin to the conveyor-belt model (Figure 12.2), which passes through every ocean in the world. Warm surface water moving diagonally northeastward across the North Atlantic eventually strikes the coasts of Great Britain and Europe and then bends towards the north; there it meets the cold polar air and ice cap and cools enough to become so dense that it sinks to the bottom.

From the ocean floor, the water streams to the south as a deep ocean current, passes beneath the Gulf Stream, and heads for the South Atlantic and, later, the Indian Ocean. With many ups and downs along the way, it meanders through the Pacific and finally finds its way back to the Atlantic, taking almost 1,000 years to complete the trip; and so the process begins again. When the current rises to the surface, it exchanges thermal energy with the atmosphere.

It does not take much to upset the conveyor-belt system at many of its intermediate points. Should the Tropics in the area of the Gulf of Mexico and the Caribbean Sea become drier than they are now, an enhanced evaporation, but not necessarily a change in the water temperature, would be the result. The water would become denser and may not have to travel as far north to cool and sink. The Gulf Stream, so important for the climate of northern Europe, might no longer reach European shores; it could even short-circuit the North Atlantic and feed directly into the deep southward current, leaving the polar region in the cold. This transition could happen suddenly. Profound climatic changes of the magnitude and quickness, caused by events such as the one just described, are known to have occurred in less than a single decade. The ice-core data, discussed in Chapter 9, are telling us that some mechanism can turn an ocean current on and off almost at will, and we have described one that could do just that. Such an event helped bring on the Little Ice Age and could provoke another.

———— 13 ————

EL NIÑO:
FROM HOAX TO MENACE

EL NIÑO IS A TERM THAT IN THE LAST FEW YEARS went from a joke to a threat taken seriously by much of the world. The authors recall when this term was the province of a handful of Peruvian fishermen and not much of anyone else. Then in 1982, and even more in 1998, El Niño and its meaning spread like an epidemic from the coast of Peru throughout the world.

The fishermen in Peru and Ecuador had noticed that about every four years the water temperature in the tropical regions of the Pacific Ocean near the coast of South America rises periodically by several degrees Celsius, while at the same time, the prevailing wind (normally from the east) slows or even reverses its direction. Heavy rainfall results and drives fish away from the coastal waters to distances unreachable by small fishing vessels with disastrous consequences that reach from Peru to as far north as Central America. Because the rains usually start around the Christmas season, some of the local people believe that they have been sent by the Christ child, or "El Niño" in Spanish. This meteorological situation carries the official name of El Niño Southern Oscillation (ENSO), although it is now known as just El Niño around the world. The El Niño of 1997–1998 was one of the most extreme of these visitations, and is certainly by far the best known. ENSO's repercussions affect the entire globe; for example, evidence indicates that arid and hot summers in the United States accompany—or more likely follow—ENSO. Other relations are also under investigation. The damaging heavy rains of the Pacific coast of North America and the heavily stagnant polluted air over southeast Asia are certainly byproducts of this now infamous intruder. Indeed, from the Arctic to the equator at least, El Niño is hated and feared, especially by those in the travel business.

For example, the boats taking tourists to and around the Galapagos Islands lose their business at these times because no one wants to make the excursion to these fantastic islands in stormy weather. The grim phenomenon leads to a nasty mess for almost everyone involved. Captain Rolf Wittmer commands a tourist vessel in the Galapagos Islands, where he was the first person of European descent to be born. El Niño is no friend of his; he looks out upon rainy islands and ocean while his business slacks off to near nothing for the year or two that the unwelcome guest plagues these volcanic islands.

Many people did not learn from the experience of 1982, the year of the strongest El Niño before 1997. Despite the research of climatologists and others, the prediction of a giant among El Niños in 1997 was greeted with ridicule from the weathermen and through them, from science as a whole. Many interviewees on the media, at least in the United States, made light of the intensive effort and technology brought by satellites and expanded computational capability. One woman interviewed on television described herself as thrilled that scientists issuing severe El Niño warnings appeared to have raised a bogus issue. But soon, the fury of the El Niño struck. California experienced rain and floods surpassing anything known up to that time. As 1997 turned into 1998, El Niño was acknowledged and blamed for everything not directly attributable to Saddam Hussein or the IRS.

A few years ago, satellite observations of the Sea Surface Temperature (SST) showed that El Niño is not just a phenomenon of local significance. The temperature distribution as measured from satellites demonstrates that it covers a large portion of the eastern equatorial Pacific Ocean, which undergoes a considerable temperature increase. This phenomenon may last for several months, sometimes for years. The wind direction, normally from the east, may reverse during the period dominated by El Niño.

Today, this scourge is known the world over. An unusually warm winter in western Europe, ice storms in Canada, inundations in several parts of the world—the 1998 El Niño caused all of that. San Francisco, the city by the bay, received so much rainfall that it became known for a while as the city *in* the bay. The reason for the onset of an ENSO, or El Niño, situation has been sought in a number of specific meteorological mechanisms. Recently, Chinese astronomers raised a new argument when they detected a relationship between the Earth's rotation rate and the ENSO events. Such a relationship is not surprising; when large air masses alter the east-west component of their motion, either changes in other parts of the world have to compensate the change or the Earth has to alter its rate of rotation. If ENSO events are related to the combined action of the solar cycle and other known periods, they might be predictable.

It was found that in many parts of the world, other meteorological phenomena occur in step with the El Niño phenomenon. The particularly hot and dry periods of the western United States is one example; another is given by a particularly dry season in Indonesia. In most cases, the weather conditions during an El Niño episode have more negative than positive aspects, and can be devastating at their worst. But another, less harmful, aspect of El Niño is related to hurricanes. During an El Niño period, hurricanes are more frequent and more severe over the Pacific Ocean; at the same time, they are less frequent and less severe over the Atlantic Ocean and Caribbean Sea.

For several decades, El Niño has been the subject of intense study. So far, the investigations have been hampered by such phenomena as volcanic eruptions, which distort the global climate and prohibit the El Niño effects from being studied in isolation. In 1997 and 1998, the situation was more favorable. The first signs of the onset of an El Niño event were observed in April 1997, six years after the Mount Pinatubo eruption, the last big volcanic event. By 1997, the effects of that outburst had disappeared, confirmed by observations of aerosols (particulate matter) at stratospheric and tropospheric levels. Thus many of the weather anomalies observed during 1997 and 1998 can be attributed solely to the El Niño phenomenon.

Under normal circumstances, the western Pacific is warmer than its eastern counterpart. A warm pool stretches from the Philippine Islands all the way to Tahiti. During an El Niño episode, this warm pool extends as far as the coasts of Ecuador and Peru. Scientists have found that the atmospheric pressures measured in the west and center of the Pacific Ocean have something to do with this phenomenon. The Southern Oscillation Index (SOI), well known and well documented for half a century, is based on the differences between the pressure anomalies measured at Darwin, in northern Australia, and at Tahiti, and is related to the presence of El Niño. The entire phenomenon comes under the ENSO, discussed earlier in the chapter. Outside the ENSO episodes, the average sea surface temperature in the warm pool is near 30°C (86°F); the eastern Pacific Ocean records an average surface temperature of around 23°C (73°F), a temperature influenced by the upwelling of cold water from the deep sea. The temperature anomalies during an ENSO are shared by waters only to a depth of about 150 meters (490 feet).

The El Niño events seem always to have been around. A record of them, however, has been available for only fifty years. ENSO events occurred during the following years: 1951, 1953, 1957–1958, 1965, 1969, 1972–1973, 1976, 1982–1983, 1986–1987, 1991–1992, and 1997–1998. They occur at irregular intervals with an average of one event every 3.6 years. The strongest ENSO events so far have been the one of 1982–1983 and, even more, the devastating event of 1997–1998.

As we have mentioned, the recent ENSO is the first in years to face no competition from atmospheric effects caused by major volcanic eruptions. The eruption of the volcano on the Caribbean island of Montserrat in 1997 has not polluted observations to a marked extent. We know that an ENSO event is triggered by a weakening or even reversal of the normally westward-blowing trade winds over the Pacific. This sets the warm pool into motion towards the east until it runs up against the coast of South America, where the sea level rises as much as 25 centimeters (about 10 inches) and the surface-water temperature rises up to 10°C (50°F). During such an episode, the entire eastern Pacific is higher than normal by some 15 centimeters (about 6 inches), based on radar observations made from satellites. This method permits the monitoring of the sea level in any ocean to an accuracy of a few millimeters.

No direct cause of the ENSO events is yet known with certainty. It is semiperiodic, but the time span from event to event undergoes large variations. The duration of the ENSO events is also quite irregular, as is the intensity. Numerous efforts have been made to predict the onset of ENSO, mostly based on mathematical expressions fitted to observations of past events. Some were successful in predicting the next event, but have failed to predict the subsequent one. Short-term predictions, however, are approaching a level of accuracy.

The last ENSO event lasted well into 1998, and by the middle of 1998, the situation returned to its normal state. A cooling of the waters off the South American coast could already be observed in February 1998, but the intensity of the devastating rain in northern Peru and in Ecuador diminished only later. The unusually dry conditions in northern Brazil and in Venezuela, which provoked extensive forest fires, continued all through March 1998. Each of the first seven months of 1998 turned out to be hotter globally than the equivalent previous record warmest month, with July 1998 being the hottest month ever recorded.

Sediments in a lake high in the equatorial Andes show an approximate four-year period of variation in the thickness and composition of the layers; these sediments were produced by swift creeks as they washed away sand and organic material. This period occurred up to 12,000 years ago, and again for the last 7,000 years; but during the interim, some 5,000 years, little or no evidence of ENSO events appears.

ENSO seems to have its counterpart in the North Atlantic Ocean, not quite as well known and not quite as disastrous. The North Atlantic Oscillation Index (NAOI) is defined as the difference between the normalized atmospheric pressures measured at Lisbon, Portugal and Stykkisholmur, Iceland. There appear to be two typical and distinct atmospheric circulations. One consists of low pressure and counterclockwise wind circulation in the region of Iceland, with higher pressure and clockwise circulation in the area of the Azores. In the opposite situation, the Iceland low is filled in and the high at the Azores reduced. The pressure difference between the two areas, the NAOI, appears to be related to the temperature of the Labrador Sea, which in turn is related to the European winter climate. Low values of the index bring cold winters to Europe and high values indicate warmer winters. Why the index switches from one mode to another is still a mystery, and although theories have been proposed, it is too soon to draw definite conclusions.

One more quasiperiodic effect is active in the Earth's atmosphere, the Quasi-Biennial Oscillation (QBO), also spelled Quasi-Biannual Oscillation, which has been detected in various meteorological situations, mostly in the stratosphere. The wind in the equatorial stratosphere changes direction every twenty-six months, and it is still a mystery as to where this strange period comes from. Its phase, east or west, seems to be related to the temperature of the vortex over the North Pole. As we will mention in the chapter about volcanism and climate, the efficiency with which aerosols, injected into the stratosphere by volcanoes in the Tropics, disperse toward higher latitudes depends on the phase of the QBO.

A QBO shows up as a variation in the direction of stratospheric winds, a correlation or autocorrelation between meteorological situations in North America and Europe; an east phase and a west phase and several other features are distinguishable. For a time scientists thought the QBO might have an influence on the depth of the ozone hole over the Antarctic. As the name indicates, the period of the QBOs fluctuates around two years, or a little less.

14

OZONE, GOOD AND BAD

OXYGEN, THE ALL-IMPORTANT ELEMENT that keeps us alive and breathing, normally occurs in diatomic form; that is, in molecules, each with two identical atoms. But under certain conditions, oxygen can combine to form triatomic molecules, where three atoms are bound together. The two forms are designated as O_2 and O_3, the second being ozone. Although ozone consists only of normal oxygen atoms, it has properties different from those of oxygen. We have seen how the so-called greenhouse gases of carbon dioxide (CO_2), water vapor (H_2O), and methane (CH_4), act to block radiation into and out of the atmosphere far above the more common diatomic molecular gases of oxygen and nitrogen, N_2, which together form about 99 percent of the atmosphere. Three or more atoms per molecule constitute the common property of greenhouse gases, with up to five atoms in the case of methane. Through their molecular construction, their interference with radiation goes far beyond that of the more abundant diatomic gases.

Ozone absorbs ultraviolet radiation. This high-energy radiation from the Sun, if not blocked aloft, is lethal to many forms of life on this planet. The modest amount of ultraviolet light that does pass to the surface causes most cases of melanoma and other skin cancers as well as the visible aging of the skin that excess exposure to sunlight causes. Ultraviolet light has also been blamed for the high incidence of blindness from cataracts, now easily replaced surgically with lens implants. Ozone must have been present in the stratosphere since organic life began to liberate the common diatomic molecular oxygen, through the decomposition of carbon dioxide in the atmosphere. Life, especially in its higher forms, has never been exposed to much ultraviolet light. It is not accustomed to it and has no tolerance for it.

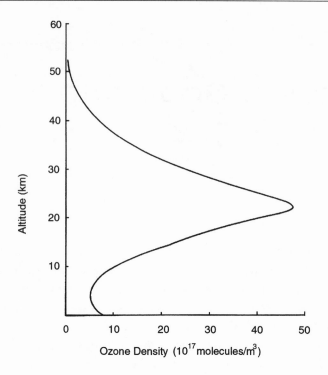

FIGURE 14.1   Average ozone concentration proportional
to altitude.

The vital layer of ozone that keeps this world habitable lies mostly at the
top of the stratosphere, its maximum concentration some 15 to 20 miles
(24 to 32 kilometers) above the ground. Figure 14.1 shows the distribution
of ozone with altitude, and reveals the peak concentration in the upper
stratosphere.

A look back to Figure 2.3 shows that our atmosphere has three distinct
temperature maxima, each at its own altitude. The lowest of these maxima
lies right at the surface; the absorption of solar energy by the surface creates the maximum. The second is created by the ozone layer as it absorbs
powerful and harmful radiation from the Sun. The third lies at the top of
the atmosphere, where the very thin air absorbs some of the extreme radiation before it ever penetrates very far into the denser air below.

All three maxima share a common property: They absorb radiation,
then heat up as a result. In between and far from them lie the two temperature minima, the lower defining the top of the tropopause and separating
it from the stratosphere; the other is farther aloft. We can see from Figure

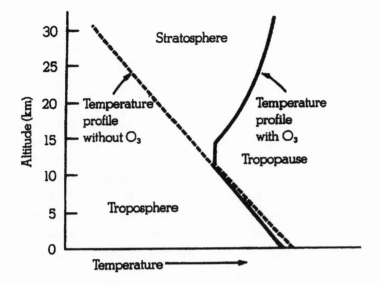

FIGURE 14.2   Temperature versus altitude with and without
ozone. The former represents actual conditions.

14.2 what the temperature of the air overhead would be if bereft of ozone;
it would be much colder up there because nothing would absorb the radiation and build up an excess of heat.

Ozone requires high energy for its formation, more than is normally
available in the troposphere; Therefore ozone is extremely scarce in the
lower layers of the atmosphere. Lightning discharges make up one of the
few natural processes in the troposphere that lead to its production. In the
upper stratosphere, the high energy of ultraviolet solar radiation is available and is completely consumed there by ozone production. As we explained in Chapter 3, free oxygen was formed early in the atmosphere's
history, and the stratospheric ozone layer must also have existed ever since.

Because the formation and the maintenance of the stratospheric ozone
layer is not yet fully understood, it has been studied extensively from various points on the Earth, including Antarctica, and from satellites. At the
British Halley Bay station in Antarctica, a considerable loss of the atmospheric ozone was first detected in October 1982; in the following months,
this "ozone hole" filled in again. The same condition repeated itself year after year, with a tendency toward increasing loss as time went on. Earlier
satellite observations, at first discarded as incorrect, were subsequently ac-

cepted as a confirmation that the ozone hole was present in the years preceding 1982. At that time, it was shown that no corresponding hole was forming over the North Pole. In spite of this difference, the situation caused great concern because it was obvious that the stratospheric ozone layer, and therefore the protection from solar ultraviolet light that it provides, could no longer be taken for granted.

The process that occurs aloft involves the formation of two molecules of ozone (O_3) from three molecules of normal oxygen (O_2) in reaction to the incoming radiation from the Sun. We have all heard that the ozone layer has been thinned and diluted in recent years as the result of human activity. Holes, relative depletions in the ozone abundance, have by now appeared above both polar regions. This discovery raised several urgent questions: What is the substance that causes ozone depletion? Why did it appear first at the South Pole? Why did it occur only during the southern spring and not at other times of the year?

It was not clear for some time after this discovery why the depletion of ozone occurred first near the South Pole and not elsewhere. Independent studies revealed that a chlorine-ozone reaction would work best at very low temperatures—around 80°C (176°F)—and on the surfaces of particles. The Antarctic stratosphere has just that temperature, but the northern polar stratosphere is several degrees warmer. At the same time, stratospheric clouds, and thus particles, are quite frequent at the South Pole but rare over the Arctic; the clouds are more often composed of nitric acid than supercooled water or ice. Nevertheless, the most recent studies show that an ozone hole of lesser extension and depth forms over the Arctic region as well. Scientists have by now observed a noticeable decline of the ozone layer over the entire globe.

Chemists had considered the possibility of reactions between chlorine and ozone that would convert ozone into normal diatomic oxygen and free the chlorine. When air samples were collected in the Antarctic stratosphere in the appropriate season, chlorine was found. Did the chlorine arise from the oceans—certain types of plankton release it there—or volcanoes, which can sometimes eject it? But not enough ozone is produced through either process, no matter the time or location. The culprit was finally identified as the chlorofluorocarbon gases, or CFCs, and known by the trade name freon (discussed in Chapter 11). They are chemically inert gases that do not react with any substance, even ozone. In all their applications, they sooner or later wind up in the air, whether intentionally or unintentionally. Their inertness allows them to linger in the air seemingly forever and without doing any harm. After their release, they travel the globe and eventually dissipate in the stratosphere. Dissipation time for CFCs is estimated somewhere between seven and twenty-six years. In the stratosphere, CFC molecules encounter incoming solar ultraviolet radiation; the radiation breaks up the molecules and frees the chlorine atoms, allowing the chain reaction to begin.

As noted, the reaction decreases with rising temperature, and is therefore most influential in the coldest air. This is why the ozone holes are limited mostly to the polar regions, and particularly to the south polar region, colder than its northern counterpart.

Thus it happens that holes of ozone depletion are observed at high latitudes, north and south, and appear to be increasing. In a later chapter on changes in health, we will discuss the adverse effects of ozone depletion such as cataracts and the rise in melanoma and other forms of skin cancer; the increases are most rapid at very southerly latitudes.

During the past decade, stratospheric ozone has been in the news many times. Experts made alarming predictions about its depletion and possible disappearance, and for quite a while arguments about the cause of the depletion went back and forth, as did the recommendations to remedy the situation. As long as scientists could not come up with a definite and overwhelmingly convincing proof that man's action was involved, an almost unbelievable reluctance arose on the part of politicians and manufacturers to adopt measures that would help preserve stratospheric ozone. This has now changed. Perhaps for the first time in history, scientists have managed to impress their findings on political and commercial decisionmakers.

After an intense struggle, international agreements were reached, including the Montreal Protocol of 1989 and the London agreement of 1991: CFC production would be phased out entirely by 1995, and most of the world is adhering to these agreements. Their effect will not be noticeable for another ten to twenty years, though, because the dissipation time mentioned above is so long. Reactions occurring in the stratosphere today are mostly from the gases released into the atmosphere in the early 1980s. Because maximum production and release was reached in the last part of the 1980s, the bulk of ozone depletion is still to come, no matter what actions are being taken now.

Recently, and partly by international agreement, CFCs with hydrogen in their makeup have replaced freon in almost all uses; the new CFCs break down in the troposphere, long before they reach the stratosphere, where they can reduce its ozone content. Even though the agreements and a different makeup for CFCs will help slow the loss of ozone, we should never-theless continuously monitor the ozone layer. The ultraviolet radiation re-ceived on the ground has been monitored for some time and, despite the stratospheric ozone loss, it has not increased as much as was expected. What seems to save us at present is the injection of manmade ozone into the lower troposphere. Ozone is a byproduct of combustion processes at a high temperature, as occurs in modern automobile engines, for instance, and by now the ozone concentration in the air sets off many of the smog alerts in California and elsewhere. Ozone is a toxic gas and is just about as undesirable as ultraviolet radiation. The ozone produced aloft is said to be good ozone; its counterpart at the surface is bad ozone.

Ozone, as noted earlier, is also an efficient absorber of infrared radia-tion. This property is not important as far as the stratosphere is con-cerned. Very little solar infrared radiation derives from sunlight because the Sun emits most of its energy in the visual region of the spectrum. In any event, almost all the infrared radiation from the ground is absorbed by the greenhouse gases in the troposphere and does not even reach the stratosphere. Within the troposphere, the contribution of ozone to the greenhouse effect is significant.

We would expect a change in the ozone to affect the weather and climate system through various mechanisms. Because the ozone present in the stratosphere absorbs solar ultraviolet radiation and thus controls the tem-perature of the stratosphere, the depletion of the good ozone would even-tually lead to a cooling of the stratosphere. Additional ultraviolet radiation leaking through would be absorbed by the troposphere or the surface, thereby enhancing the greenhouse effect. Although the troposphere would end up warmer than at present, the stratosphere would be cooler. Just what this increased temperature contrast can do to the climate system as a whole is yet unknown. The situation is being taken into account in the cli-mate models, which we will discuss later.

Measurements of solar ultraviolet radiation are made from the ground, from balloons, and from satellites. Direct measurements of the ozone con-tent are being made from commercial aircraft. Recent observations show that ozone depletion now reaches altitudes up to 12 miles (20 kilometers), which is somewhat higher than it was in 1995. It is now estimated that CFCs take about five years to reach the bottom of the ozone layer. Stratos-pheric ozone is presently expected to reach a minimum soon after the year 2000 and to begin recovering by 2010. Ultraviolet radiation is presently increasing at rates of 7 percent and 10 percent per decade at the latitudes of 55 degrees north and south, respectively.

Although rocket exhaust plumes also cause a loss of good ozone, their contribution to the total depletion of this gas is insignificant. Supersonic aircraft such as the Concorde may create a greater problem should they become numerous because the sulfuric acid they leave in their wake may contribute to the ozone loss.

Measurements of the atmosphere's total ozone content are made by observing solar ultraviolet radiation after it has passed through the atmosphere. This means that near the poles, no measurements are possible during each pole's winter season because the Sun remains below the horizon. Nevertheless, from latitudes of less than 65 degrees north and south, measurements can be made throughout the entire year because some scattered solar light is available at all times. These observations show that ozone depletion is detected already by the middle of the winter when no solar light is available at the pole; therefore, the entire mechanism that leads to ozone loss may require rethinking.

TWO WORLDS

No man is an island, entire of itself; every man is a piece of the continent, a part of the main; if a clod be washed away by the sea, Europe is the less, as well as if a promontory were, as well as if a manor of thy friends or of thine own were;

Any man's death diminishes me, because I am involved in mankind; and therefore never send to know for whom the bell tolls; it tolls for thee.

JOHN DONNE, *DEVOTIONS XVII*
ERNEST HEMINGWAY, *FOR WHOM THE BELL TOLLS*

IN ADDRESSING THE GLOBAL WARMING DILEMMA, it is not our intention to define the first and third worlds; these terms are well understood, even now when the second, the alleged communist world, has largely disappeared. It is generally understood that the differences between these two remaining worlds lie in their standards of living, particularly average personal income. One of the authors resides in the first world, the other in the third. For two reasons, people in the third world look upon weather and climate differently from those in highly developed nations. First, the two worlds are for the most part on a north-south divide: Most underdeveloped countries are near the equator and have tropical climates. The second reason, just as important, is that people of low income live closer to nature. Their houses are less sophisticated than those in developed nations and far more vulnerable to the effects of changing weather than are solid, well-equipped houses. Because people in the third world travel more frequently on foot or on animals rather than by car, they are again more exposed to weather conditions. Consider the difference between life in the big city and on the farm: New Yorkers rarely note the weather unless they need to carry an umbrella, and they almost never look at the sky; the country person lives with his sky. In any event, newspapers print the forecast on the front page with details inside, and television stations provide regular weather reports.

These realities lead to a closer observation of weather abnormalities, of which the most prominent receive specific names. The best-known example of this is the infamous El Niño, discussed in Chapter 13. Fishermen along the coasts of Colombia, Ecuador, and Peru first noticed and named El Niño. They suffered devastating setbacks to their livelihoods each time an El Niño struck.

In tropical lands, the seasons are difficult to differentiate because of the slight temperature variation between them—if one exists at all. Even so, the terms *summer* and *winter* are commonly used in Venezuela and other tropical countries, but their meanings in those countries have more to do with rainfall than with temperature. Rainy periods are called winter; sunny ones, summer. A European or North American visitor who experiences the two seasons in the same day may not think of them as seasons. The ratio between annual and diurnal variation defines one of the major definitions of tropical and temperate climates: The diurnal is of greater amplitude than the annual in temperature near the equator, and particularly in desert regions, where the daily temperature range is greatest.

Venezuela is among the tropical countries that such violent weather as tornadoes and hurricanes rarely touches. Wind speeds of more than about 40 miles (65 kilometers) per hour are rare; indeed, the flimsy roof construction, which would never withstand a major storm, verifies this. Some coastal areas in the Tropics do suffer the onslaughts of hurricanes and typhoons. The coast of Bangladesh, facing the Bay of Bengal, consists of many low-lying islands hardly above sea level, and consequently of no protection against storms. The sea swell alone can take many lives, and many of the major disasters of the twentieth century involving loss of life occurred on the Bangladesh coastal plains. Of the nine natural disasters in the last century that have each taken more than 100,000 lives, two resulted from tropical storms striking Bangladesh; three from floods in China; and four from earthquakes, three in China and one in Japan. The six most lethal storms took place in the Bay of Bengal.

A solution to the global warming problem must be global. The first world has made some moves to lessen the impact of disaster, the most important being the control of population growth. Much of the first world's

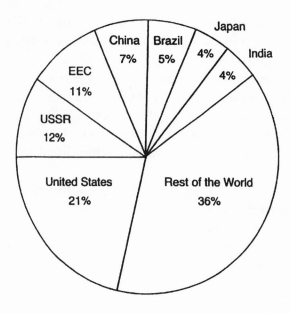

FIGURE 15.1   Contributions to global warming in the 1980s by country (percent).

growth is now dependent upon immigration from underdeveloped societies.

Although the largest of the developed nations, the United States, has less than 5 percent of the world's population, it consumes more than one fifth of manmade consumption of fossil fuels (which spew greenhouse gases into the atmosphere) as indicated in Figure 15.1. The U.S. Gross National Product (GNP), the sum of all goods and services, reflects the inequity because it forms a large part of the world's GNP. Nevertheless, the United States and other developed countries of the world share in the emission of gases fouling our air to such an extent that they are obliged to make efforts to reduce them. But what can we say about the methods these countries use?

Let us consider present energy demand and its probable future development. The United States has the highest energy consumption per capita, taking into account all energy consumers, closely followed by western and central European countries. It is true that from about 1980 the energy efficiency of transportation, domestic devices, and industrial machinery has increased remarkably. Even so, we have not yet reached the peak of energy consumption per capita.

The third world presents a different picture. There we should expect an almost explosive increase of energy demand in the future for two reasons. The standard of living ranges from low to practically absent, but the third world has every reason to aspire to the basic commodities of modern everyday life in the industrialized world. We are talking about electric light, refrigerators, air conditioning (heat will hardly be required because the third world is almost confined to the Tropics), and efficient and comfortable means of personal and public transportation. To this we must add rapidly increasing populations; low standard of living and lack or absence of efficient birth control are natural consequences of the economic situation. An improvement of the third world's economy coupled with widely available birth control methods and employment for women are in the interest of the entire world.

No one in richer countries has the moral right to consign the people of the underdeveloped world to eternal poverty. People in poor countries also want refrigerators, good lighting in their homes, air conditioning, telephones, television, and, above all, automobiles—commodities that in the developed world are considered essential.

"Business as usual" in the developed nations involves practices that unacceptably waste energy. To convert their own excess production into financial gain, the industrialized countries export energy-inefficient and long-outdated or even obsolete household appliances and automobile engines to the third world. In the United States, automobiles have been required to carry catalytic converters, and gasoline stations to market unleaded fuel, since about 1975. Venezuela began to do the same only in 1999; unleaded gas has been unavailable there. Nevertheless, automobiles using leaded gasoline are still being exported to Venezuela.

Planned obsolescence means that consumer goods have a built-in lifespan: After a certain time, they must be replaced. Such tactics are particularly noticeable in the third world. Why must automobile tires manufactured in Venezuela last for less than half the mileage of those made in the United States? Why must Venezuelan refrigerators rust from the inside out (because no anticorrosive paint was applied to the inside of the metal case) and have to be replaced after five years? Why do light bulbs have a notoriously short life in South America? How much energy must be spent to produce more refrigerators, tires, and light bulbs? How much extra trash fills the junk yards and garbage dumps? How much energy does the hauling of all this trash consume?

Countries with rapidly expanding populations will not achieve a much higher living standard soon. But China with its lower birth rate still has almost five times the population of the United States. What happens to the atmosphere should that huge but relatively stable population acquire the material benefits of the good life? Can the world survive a billion cars and trucks driving on the roads of China? Where do we draw the line between the "haves" and "have nots" in the future, especially if polluting devices must be rationed for the Earth to survive? And how and at what point do we force our largest industrial nations and economies to abandon "business as usual"? "Blessed are the peacemakers: for they shall be called the children of God" was written in a time faced with different problems. Blessed will they be now; for we need them.

These tough questions were raised at the Kyoto Meeting in Japan, a conference of delegates from more than 150 nations that met in December 1997 to discuss greenhouse gas emissions. There are no easy answers, but with global warming threatening us all, we must try to find solutions, even if it means abandoning nationalism as we know it.

The burning of the rain forests in Brazil and Venezuela, in Madagascar, and in Indonesia injects a considerable amount of extra carbon dioxide into the atmosphere. We might think that if the forests grow back in several decades they will take the same amount of carbon dioxide from the air again, and in the long run nothing will have changed. This is not the case. In the Brazilian rain forest, farmers exhaust the soil of the rain forests they burn, a process we described earlier in the book; it may therefore take thousands of years for the forest to replace itself. More than laws are needed to stop this activity. The government, maybe with the aid of the rest of the world, must help these farmers find a way to subsist without endangering the rain forest.

In northern Brazil and southern Venezuela, the situation is even worse. As noted in Chapter 11, Brazilian gold miners burn large tracts of forests and wash away the scarce topsoil to reach gold-bearing sand. To wash out the gold, the miners divert the rivers, load them with mercury, and poison all river flora and fauna downstream. When the Yanomamis (the indigenous people of Brazil) try to protect their very existence by resisting the miners, they are driven out or killed.

Indonesian farmers use the same techniques as Brazilian farmers, but, as we discussed earlier, another factor in Indonesia also contributes to the destruction of the forests: the harvesting of precious wood. Large strips of forest are cleared just to reach and cut down a giant tree more than 1,000 years old. Such destruction produces fancy furniture for the well-to-do in distant lands—people who could have had the same furniture made from homegrown wood at a fraction of the cost, thereby saving virgin rain forests.

It is estimated that at the present rate of burning virgin forest, carbon dioxide emissions into the atmosphere amount to something like 7 percent of the total global human carbon-dioxide input. This does not sound like much and it will soon end anyway, when all the rain forests are gone.

16

ENERGY: PRODUCTION
AND CONSUMPTION

The key point here is that the Netherlands, Bermuda, and
Monaco (and our very large cities) can be crowded with
people only because the rest of the world is not.

PAUL EHRLICH

THESE ARE THE HUMAN ACTIVITIES that most adversely affect the
Earth's atmosphere:

1. The production of electrical energy with coal, oil, or gas-burning
   engines.
2. The use of combustion engines for ground, air, and marine
   transportation.
3. The primary production of steel with iron ore and coal.
4. Burning vegetation.
5. Specific industrial production processes.

We will take a closer look at each one of them, but first let us consider
present energy demands and probable future development. Leaving aside
transportation, which we will look at elsewhere, the most widely used
form of energy is electrical energy. Electrical energy has an extraordinary
advantage over all other forms of energy: It can be transported over great
distances with little loss and is easily distributed to an almost limitless
number of outlets. Much of the electrical energy produced today comes
from plants powered by coal-, oil-, or gas-driven engines. In all cases, combustion releases carbon dioxide into the air along with such contaminants
as oxides of nitrogen and sulfate. Among these fuels, natural gas is the least

contaminating. Two ways to reduce the contamination caused by the production of electrical energy are (1) use alternative energy sources not based on combustion; and (2) reduce the use of energy. Many alternatives are available, and most of these are already being adopted in some regions of the world.

At this point, we want to show how energy consumption differs widely from one country to another. More important, though, is the production of atmospheric contaminants, which is closely related to energy production and consumption. We can use the yearly total injection of carbon into the air per capita as a measure, including not only domestic and personal contributions but also transportation, personal and public, and all industrial uses. Keep in mind that the countries with the largest numbers produce many of the industrial goods that are exported to countries with low carbon injection.

Atomic power is the best-known and most widely applied alternative energy source. No doubt atomic power is clean in the sense that it does not contaminate the air as long as the nuclear power plant is operating properly. But accidents happen, and in 1986, Chernobyl, the most infamous, produced an invisible and deadly contamination that reached many parts of Europe in 1986. Disposing of the residuals left by the burning process also poses a serious and ever-growing problem.

Certain countries, among them the United States, have already stopped building new atomic power plants. Other countries, France being the leader, produce most of their energy in atomic plants. Some countries, for instance, Argentina and Brazil, are just now entering the "atomic age." Although there has been no major breakthrough among other methods of energy production, it is doubtful that atomic power will be the energy source of the future.

Solar energy is an oft-quoted alternative, and it can be used in two ways: Solar radiation can be concentrated, or "bundled" by such optical means as a parabolic mirror; the concentrated radiation then passes through a heat-absorbing element that transfers the heat to water. The resulting steam can drive a generator. But the second method, the so-called solar cells, is more practical, and simpler to use. Solar cells transform radiation directly into an electrical current, which can then charge a battery. Solar cells are presently the most popular way of using solar energy, principally because the method is simple and can be installed in all sizes. Today, many homes in Europe and the United States draw most or all of their electrical needs from a "solar roof."

An interesting pilot project for large-scale energy production is being tested in Spain. A huge upside-down transparent funnel of about 1 square kilometer is placed over a flat surface. The idea is that the air under the funnel heats and rises, its only way of escape being through the opening of the funnel at the top. At the opening, the heated air, now a strong air current, drives a large propeller, which in turn is connected to a generator.

All solar devices work only during the day, and reach full efficiency when the sun is high. When the sky is overcast, they are inefficient, although not totally dead. Complete dependence on solar power requires high-capacity energy storage, most easily accomplished with batteries. But there are other ways of storing energy. One nonconventional form has been adopted in Hamburg, Germany. During times of low energy demand, the existing and idling generators provide the required energy to pump water into a basin above the ground. This water is then released through a hydroelectric power plant during times of high energy demand.

As attractive as solar energy may appear, it is not the final solution to the global energy demand. For a city such as Caracas, situated in the Tropics and with sunshine all year, it is impossible to capture enough energy to meet all electrical energy needs. With present-day means, solar radiation would have to be captured over a surface of about 1,000 square kilometers (620 square miles), but not enough solar radiation reaches the Earth to accomplish this. It has been suggested that the Sahara desert could be converted into an enormous power plant supplying the world with electrical energy. Unfortunately, such a solution is as yet impossible in view of the world's political structures. It is also unfortunate that the world's oil reserves are unevenly distributed.

Hydroelectric energy as an alternative for the generation of electricity can be applied efficiently in mountainous areas having generous precipitation. The largest power plants in the world are driven by water power generated at the Hoover Dam in the United States, the Guri Dam in Venezuela, and elsewhere. These will soon be surpassed by the Parana River Dam presently being built between Brazil and Paraguay. Another giant is the Aswan Dam in Egypt—proof that taming water to generate electrical energy is not always beneficial. For thousands of years, Egypt's economy depended on the yearly flooding of large tracts of land by the Nile River. Not only did the Nile bring adequate irrigation water but it also supplied a well-balanced fertilizer through its mud. Now that fertilizer has sunk to the bottom of the lake behind the dam. Today, Egyptian farmers use artificial fertilizer and depend on artificial flooding. The Nile delta was once one of the most fertile regions of the world, built up and maintained by the mud the Nile River carried to the Mediterranean Sea. Now the Nile flows into the Mediterranean without its mud; as a result, the sea has rapidly eaten away at the delta.

Because particles carried in flowing water settle to the bottom when they reach still water, artificial dams create silt; this fills and clogs their reservoirs, reducing their capacity to store water. How long it takes for a reservoir to become useless depends on how much silt contributing rivers and creeks carry, which, in turn, depends on how much of the surrounding surface is eroded by rain. Surface erosion depends on how well vegetation protects the surface against erosion, and the condition of vegetation depends on the climate. The presence or absence of forests in a given region has an important effect on the local climate, which is one more reason to preserve the forests in North and South America at any cost.

We have just shown why hydroelectric power plants can be expected to have a limited useful lifetime. But accumulated silt can be excavated and removed and the original capacity of a lake restored, although this is prohibitively expensive. Even so, it has been done in places; for instance, the Santo Domingo Dam in Venezuela had its silt removed because the dam also serves to control flooding.

Tidal waves, created by the gravitational action of the Sun and the Moon, contain an enormous amount of energy. In open ocean waters, the tidal range (the difference in level between high and low tide), is only about half a meter. However, when east-traveling tidal waves reach an obstacle, a lot of water can pile up. Along the Atlantic coasts of Britain and France, the tidal range can reach 24 meters (about 78 feet). At high tide, reservoirs could be filled; at low tide, water could be released into the low sea to drive turbine generators. But such an electric-power-generating system could work efficiently only in a few parts of the world. Because of intermittent energy production, tidal power plants would require electrical storage capacity. But if they contributed to a network of power stations, they could relieve the load on other generators during the times they were able to produce.

The wind-driven ocean waves contain so much energy that floating devices have been designed to make use of the differential energy contained between the high and low parts of waves. This would be an interesting solution for a coastal area where permanently high waves could be found not too far offshore and where such an installation would not endanger marine traffic.

Electric generators powered by wind have become popular since about 1980. Giant propellers with three arms are now a common site in Denmark and northern Germany, and California and southern Spain have hundreds of them. Many farms in Europe now draw most of their power demand from their own "windmills," which remain connected to the general power network and draw electricity from it when their demand is high and their own production insufficient. In many cases, the power companies buy their overproduction when the wind is strong and steady and the local demand is low. Again, wind-powered electricity generation is intermittent and requires either storage capacity or a combination with other sources of energy. Electric generators are efficient only where the wind is strong and frequent.

The internal heat of the Earth is for practical purposes an infinite source of energy of which little use has so far been made. As any miner can testify, the Earth gets warmer as one descends through a shaft. The normal increase of temperature, or *geothermal gradient,* is about 1°C (2°F) for every 30 meters (about 98 feet) in depth. The geothermal gradient is not uniform over the Earth. There are *geothermal regions* with a gradient as low as only 5 meters (16 feet) for 1 degree. Regions with hot springs always have a steep geothermal gradient, but there may also be regions with steep gradients without hot springs. Iceland and New Zealand make use of their many hot springs, both countries employing the heat underground for energy production.

A geothermal power plant works on a simple principle. Two holes with large diameters and in close proximity are drilled to a depth where the rock reaches a certain temperature. Below ground, the two holes are connected by a large cave filled with crushed rock. When one of the holes is filled with cold water, the water flows into the cave where it heats and, because of the pressure difference, rises to the surface through the other hole. Because hot water is lighter than cold water, the water spouts from the second hole under pressure. When the water is hot enough, it emerges as steam, and can then be used to drive a steam turbine coupled to a generator.

Geothermal power plants can, in principle, be built and operated anywhere because underground heat is found everywhere. These plants are expensive to operate, however, and the deeper the drilling, the more expensive they become. Although geothermal power plants could be built in cities, costs could be reduced by building them in regions with favorable geothermal gradients; high-tension lines could then transport the electricity to the cities, but also at high cost.

Geothermal power plants are so new that not much information exists concerning the kinds of problems they might generate. It is known that the two holes and the cave require considerable maintenance and must be kept free of debris. But geothermal plants could be an option for countries without oil, coal, or mountains.

Wood has been the primary source of energy for mankind for thousands of years, and in many places it still is. When we burn scrap wood in the fireplace, we save on our electricity and oil bills. If not burned, scrap wood decomposes on the garbage heap; there, microbes convert the wood back into carbon dioxide, which then escapes into the air. By burning wood, we simply short-circuit this natural cycle, but on the whole do no damage to the atmosphere. If we use the ashes as fertilizer, no real damage is done anywhere, but energy has been saved elsewhere. Wood burning is done on a grand scale in Finland, where power plants are fired by wood from forests owned by the power companies. Because trees are constantly replanted, no new forests are cut down. Finland, with its many woods and small population, can afford such a scheme. However, because supplying a medium-sized city with electrical energy derived from wood requires an excessively large forested area, this method is unlikely to be copied elsewhere.

When organic matter decomposes, the methane produced contributes to the greenhouse effect. Methane is burnable, even explosive when mixed with the right amount of oxygen. A garbage dump can produce huge quantities of methane, and some years ago, a garbage dump in England exploded, killing several people and seriously damaging nearby houses. Some major cities are already harvesting methane from garbage dumps and using it for industrial and domestic purposes, thereby taking advantage of cheap energy and reducing the amounts of methane escaping into the atmosphere.

Cattle manure, another methane-producer, supplies farmers with fuel for cooking and heating; many farms in Europe and North America now have methane-extracting installations. Methane combustion produces mostly water vapor with some carbon dioxide.

On a global scale, fossil fuels (coal, oil, and natural gas) are still the primary sources of energy production. France depends upon atomic power, and Switzerland, New Zealand, and Iceland, hydroelectric power. As we have shown, most of these methods are contaminant. Reducing energy use is one obvious step towards reducing atmospheric contamination and loading with greenhouse gases; using energy more efficiently is another. Let us look at both possibilities.

Because the burning of fossil fuels threatens the future of the Earth's atmosphere, a more efficient use of fossil fuel can help protect the atmosphere. First, cars can be smaller. European and Japanese manufacturers have long built small four-passenger cars with engines big enough to keep them moving at a reasonable speed almost anywhere. American manufacturers are now following the trend. Much progress in building fuel-efficient engines has also been made in recent years. In spite of the measures already adopted in many parts of the world, the total number of cars on the road and total worldwide fuel consumption are increasing, even in highly industrialized countries; therefore, we need a basic change in fuel used for transport. At present, only oil or coal derivatives are practical and efficient power sources for automobiles. If we want to use hydroelectric or geothermal primary energy to power our cars, we must use these energy sources to produce some other kind of portable energy. Batteries are an obvious choice, but there are other possibilities. Electric cars have been around for a long time, but only for specialized purposes. For delivering packages, the German postal service uses electric cars because they are more adaptable to frequent stops and starts, but these vehicles were never meant to compete with the common automobile. The development of electric cars for everyday use began several decades ago, and some of the large automobile manufacturers now have electric cars on the market. The electric car has a number of important advantages over vehicles with combustion engines. Here are some of them, but not all may apply to all models:

1. They do not contaminate.
2. They do not consume energy while the car is idling.
3. They are almost noiseless.
4. They recover energy when decelerating by switching the motors to the generator mode.
5. They are natural four-wheel drive vehicles, having one motor on each wheel.
6. They use natural power steering through a change in the voltage difference between left and right wheels.
7. Batteries can be charged at home or in parking lots.

Electric cars have two primary disadvantages:

1. Their autonomy, their maximum speed, and their acceleration time to a given speed fall behind those of conventional cars. We

must remember, however, that today's electric cars are the product of only a few years' research and experience, whereas the conventional car is the result of more than a century of practical experience.

2. They are expensive, but future mass production will make them more affordable.

Scientists and researchers have given a lot of thought to fuels that don't contaminate the air. Hydrogen-burning motors show great promise. Because hydrogen at room temperature is a highly explosive gas and extreme precautions must be taken when handling it, compressed and solid hydrogen are unlikely to become the fuel of the future. Hydrogen-loaded metal blocks that can release their hydrogen content on demand have been developed in recent years, and although they sound like a feasible alternative, autonomy (mileage from a full tank) must yet be evaluated.

Charged batteries and metal blocks full of hydrogen must be manufactured, which means using electricity generated by a power plant. It may seem wasteful to use a primary power source's electrical energy to charge batteries or blocks that drive a car; after all, we might as well use the primary energy for the car. But because a combustion engine must run at various speeds and push both light and heavy loads, it cannot be built to work at optimum fuel efficiency all the time. A combustion engine at a power plant, however, always operates in optimum conditions; furthermore, converting electrical energy into a charge in a battery or into hydrogen in a block is efficient from the beginning. In turn, converting a charge in a battery into the motion of a car is efficient. Therefore, although fossil fuels are still the main primary energy, it makes sense on a global scale to change to electric cars—unless something even better is invented.

Of all the industrial processes, two have a particularly strong influence on the atmosphere, namely, the production of steel and aluminum. Iron ore and coal are burned together, and their types must match well to produce high-quality steel. For example, iron ore from northern Sweden is matched perfectly with coal from western Germany. Of all fossil fuels, coal pollutes the most, but there is no other way to make steel. To reduce atmospheric pollution and the emission of carbon dioxide, we must reduce worldwide steel consumption. In Europe, a number of steel mills have closed since about 1980 (although the reasons may not be consideration for the Earth's atmosphere). Although recycling scrap iron helps reduce pollution, high-quality steel cannot be made from scrap iron only, and the amount of scrap that can be mixed with new steel is limited.

Aluminum is made from bauxite in a process that requires a huge amount of electricity. Used aluminum can be fully recovered, thus its recycling is profitable in many ways and helps reduce the consumption of electricity. Glass, too, requires energy for its production or manufacture. Melting existing glass takes less energy than making glass from natural quartz. Recycling bottles is better than remaking them by melting recycled glass, which, in turn, is better than making them from scratch. Above all, the consumer dictates how much of any product is made. By buying milk or beer in bottles that can be returned, by using metal objects for as long as possible, and by recycling all waste that can be recycled, anyone can help conserve the Earth's atmosphere.

Garbage is another important source of energy consumption. If we account only for homes and offices, the per capita production of refuse in the industrialized world exceeds a kilogram (two pounds) every day. Most of this is packing material and throw-away plates, cups, knives, forks, etc. All of it is produced somewhere through an energy-consuming process and transported to various outlets. Later, this refuse is transported to garbage dumps, where it is handled, disposed of, buried, or whatever. If we could cut our waste by half, only half as many garbage trucks might be needed. Proper domestic waste separation into types of waste products results in more efficient recycling of certain common materials and also saves energy. Furthermore, do we really need the many advertising inserts that arrive with the newspaper every day? Might consumers elect whether or not to receive them?

Germany has adopted a law that helps reduce wasteful production: Stores must now take back unwanted packing material from the consumer at the moment of purchase. This measure has put such pressure on retail outlets that industry has been forced to use less packing material.

Industries have a natural interest in reducing their energy consumption as long as it reduces their total production costs; if they accept a cost increase, they pass it on to customers. This means that some of the burden for the measures industry takes to reduce energy consumption falls, directly or indirectly, on the consuming public.

Automobiles, trucks, and airplanes are some of the leading consumers of oil and oil products. A superficial look at a traffic jam on a highway or at the vehicles piled up behind a stoplight shows that idling engines waste an enormous amount of fuel. On top of that, most cars are occupied by only the driver. We can take a number of measures to reduce fuel consumption significantly. Let us look at some already adopted in various parts of the world.

Outdated methods of traffic regulation are among the least conserving of human activities. Traffic signals have become articles of faith, often for no reason. When an accident occurs, especially in a residential area, a clamor rises for a traffic signal where none is really needed. Of those signals necessary for rush hour, many continue their cycles twenty-four hours a day, causing car and truck engines to idle alone and needlessly. The average automobile is reported to require an extra gallon of gasoline for every six traffic signals. Six red lights require a gallon more than the same six lights in flashing mode; for trucks, the difference is far greater.

In the 1980s, Connecticut removed toll gates on all roads and bridges within its borders, and because traffic had increased so much over the years, the fuel savings were considerable. But in a move backwards, Connecticut exempts vehicles over five tons—the major polluters on the road—from the statewide automobile emissions inspection program. Fuel will be saved when these problems are corrected.

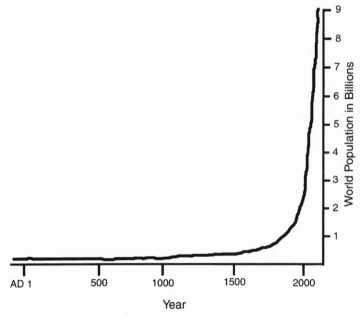

FIGURE 16.1   Population increases from A.D. 1 to 1992 and projected to 2032. The population explosion of the twentieth century is obvious.

As the first world must rethink its attitudes towards energy, so must the third world come to grips with excessive population growth. As Paul and Anne Ehrlich state in their recent book, *Betrayal of Science and Reason*, "Whatever your cause, it's a lost cause if humanity doesn't solve its population problem." The first world earns high grades here only comparatively, and considerable progress is now noticeable in Latin America and on the Indian subcontinent; in Africa, population growth rates are still high—about 3 percent per annum—making it the second most populated continent. As women seek and obtain jobs in the underdeveloped countries, family planning becomes a necessity and the growth rate begins to fall. Figure 16.1 addresses this greatest of all problems.

17

THE WORLD IN A
FUTURE CLIMATE

WE HAVE SEEN CLIMATIC CHANGES that confirm an overall warming trend. We gained a much greater understanding of past temperatures with core samples of ice as much as 500,000 years old extracted from the Greenland and Antarctic ice caps. Scientists are now able to say with conviction that summers over the last decade or two have been among the hottest in 500,000 years! Since 1980, North America and Europe alone have experienced heat waves of unprecedented severity and length. If the ten most extreme cases of anything were found in the last twenty among many thousands of cases, the laws of probability would not permit us to see this as random.

We have known for years that carbon dioxide is the prime cause of greenhouse warming in the atmosphere. The gas is an efficient blocker of infrared radiation, more effective than the more abundant gases (nitrogen, oxygen, argon, and water vapor) combined. For centuries, carbon dioxide remained at a level near 0.028 percent, equivalent to 280 particles per million air particles (abbreviated ppm) in the total atmosphere. At the start of the Industrial Revolution, this level began to rise slowly, from 280 ppm to about 300 ppm by 1950, and then more rapidly to near 370 ppm today. The steep rise in the abundance of this gas has been closely monitored since 1958. The amount varies annually, rising with the disappearance of foliage in the Northern Hemisphere each fall and winter and declining with its emergence in the springtime. But a steady overall increase is also present; both the annual variation and the steady rise appear in the curve shown in Figure 1.1. The increase is the result of greater human consumption of fossil fuels, and lately of increasing deforestation, especially in tropical regions. Carbon dioxide leads the way, but methane and other gases also act as greenhouse gases and they, too, are on the increase.

Does human activity, in particular our input of greenhouse gases and aerosols into the atmosphere, affect the future of global climate? This question is of great interest and concern worldwide. Behind this question looms another of maybe even more importance: Does it really matter if the global temperature rises by a few degrees in the foreseeable future? Depending on the answers to these questions, drastic steps may have to be taken worldwide to avoid a global climate catastrophe; alternatively, it may be that we can continue with our present rates of energy consumption and population growth with no serious negative effects for decades to come. Any scientific reasoning or speculation about the future climate and its consequences in the various parts of the world should answer these questions: Which trend will the global temperature follow during the next half century? What are the expected effects of a climate shift on human society and on all other forms of life? Answers to these questions will determine which actions must be taken by individuals, communities, nations, or all of mankind. The actions may be costly and they may seriously affect the economies of some nations, and thus will meet resistance.

Let us look first at the future of global temperature. The answer or answers to our first question can be found only from model calculations as outlined in an earlier chapter. At that time, we pointed out that the predictions are still uncertain, but we can be confident in the predicted reaction of the Earth's atmosphere to various amounts of human inputs. We have already shown (see Figure 9.1) how carbon dioxide has increased since the beginning of the Industrial Revolution in the nineteenth century. No doubt this increase is caused by the burning of fossil fuels. Other greenhouse gases, for instance, methane, follow a similar trend, and for the same reason. What will these inputs be in the future? It is fantasy to expect human inputs to halt so that the atmosphere has a chance to recover. And it is as unrealistic to expect a substantial decrease in greenhouse gasses escaping into the atmosphere in view of undeveloped nations' efforts to improve their living standards and their difficulties in controlling their populations; the unwillingness of developed nations to sacrifice part of their comfort compounds the problems. Thus we must base all climate projections into the future on an ever-increasing amount of greenhouse gases. Attempts have been made to curb gas emissions; at the international conference in Kyoto in December of 1997 (discussed in Chapter 15), a general reduction of something like 15 percent over the following several years was proposed. The representatives of most participant nations signed the respective commitments.

Some time during the twenty-first century, carbon dioxide (along with aerosols and other greenhouse gases) in the air will double from its level at the advent of the Industrial Revolution in the middle of the nineteenth century. Climate modelers calculate the behavior of global temperatures through various assumptions about the rate of greenhouse gas increases. Although the modelers come up with various curves, they all agree on the trend, namely, temperatures will increase globally and in proportion to injection rates.

What will be the effect of a global temperature increase of a few degrees? This depends on location. In general, climate zones will shift towards the poles, and uphill in mountainous areas. Thus the Tropics will expand into what are now the subtropics, and these will push into the temperate latitudes, forcing them into the polar latitudes. Trees will grow at higher altitudes along mountain slopes.

Canada may expect to enjoy a less frigid climate and the Southern states could turn into deserts. Northern Europe and Siberia will become agricultural, and such Mediterranean countries as Spain, Italy, and Greece, and all of North Africa, will have difficulty maintaining their food production. Deserts will also expand in India and China. Changes expected for the Tropics will be minor because the temperature increase there will be minimal.

A rise in sea level does not just inundate and flood beaches. Most of the damage is caused by greater wave action as it scours sand from the beaches. The Bahamas and the eastern seaboard of the United States might erode away to nothing in a few decades; indeed, local hurricane damage is a foretaste of what could lie ahead.

Of great concern is the fate of the ice on Greenland and Antarctica and of Antarctica's grounded ice. Should all this ice melt, thereby flooding the global ocean system with huge amounts of water, the worldwide sea level is expected to rise by about 60 meters (almost 200 feet). It is doubtful that this will ever happen, and if it did, it would take a long time. A global warm-up means that water will evaporate at a higher rate and more vapor will circulate through the air, the resulting increased precipitation falling in the polar regions. But if all the ice melted, most of Belgium, practically all of the Netherlands, a good part of northern Germany, and most of Denmark would be under water, and so would Venice and parts of cities such as Bangkok, Miami, Tokyo, and many other densely populated areas.

The first to disappear from the inhabited world would be a number of low-lying islands in the Pacific and the Indian Ocean. In fact, these already feel the threat. A rise in global temperature could indeed cause the polar ice to disappear if not for the precipitation that would partly or even fully offset the melting. A warmer atmosphere and warmer water would bring about a higher rate of evaporation and cloud formation, and precipitation would increase as a consequence. But just where the additional rain or snow will fall is uncertain. Could it be at the poles? This is a possibility, and it could well be that in spite of the higher temperature and the faster melting rate at the poles, the accumulation of new snow may be even faster, causing the polar ice to grow. Although climate models predict that the polar ice will decline, the calculations have margins for error.

The latest observations indicate a trend. The open waters that form every summer amid the floating ice on the Arctic Ocean are becoming wider every year. There are also indications that the ice around Antarctica has been receding during the last couple of centuries, and that this retreat is still occurring. Unfortunately, there is no way yet to prove that the re-treating ice is caused by manmade global warming.

The Intergovernmental Panel on Climate Change (IPCC), mentioned in Chapter 10, sets the rise in ocean levels over the past century from 4 to 10 inches (10 to 25 centimeters). Glaciers melting in Alaska seems to be the major contributor to this increase. Alaska comes in fourth (after Antarctica, Greenland, and the island archipelago in extreme northern Canada) for amounts of glacier ice and ice pack on land.

The input of fresh water from rain and rivers into the oceans acts to lower their salinity and therefore their density; this is one of the motors that keep the ocean currents going. A major change in the precipitation may cause a change in the course of these currents. It is conceivable that the Gulf Current, a major factor in the climate of Europe, may flow elsewhere or cease to flow at all as an indirect consequence of global warming, resulting in Europe's experiencing a colder climate in the event of global warming; such an event is unlikely to happen, but the possibility exists that global warming could cause parts of the Earth to cool.

Although more water vapor will be pumped through the air in a warmer climate, and, in turn, more precipitation, the local picture may be quite different; indeed, some trends are already emerging. In 1998, Mexico, Nicaragua, Honduras, Peru, Ecuador, and several other nearby countries reported heavy rainfalls and catastrophic flooding. At the same time, winter in Chile brought almost no rain or snow, leaving the reservoirs dry, threatening major cities with a serious water shortage, and denying agriculture its desperately needed irrigation. Most of Chile depends on winter rainfall because summers are usually completely dry.

It is possible that a point of no return could be reached where heat continues to escalate, no matter what we do. We have only to look at Venus to remind ourselves of a greenhouse disaster. As we discussed earlier, our lovely evening star has a "runaway" greenhouse effect, with carbon dioxide heating its surface to some 800°F (450°C); this far exceeds temperatures the Sun alone would produce there.

Naturally produced carbon dioxide has heated the Earth some 30°C (54°F) above the temperature it would have maintained without this gas. But until now, the temperature of our world has been in equilibrium. Homo sapiens is performing a grand unplanned experiment to discover the effects of a doubling or more in the abundance of greenhouse gases. The answer will come at a fearful cost.

The average rise in sea level is 1.5 millimeters (0.06 inch) per year; this has been observed over the past century along the coastlines in many parts of the world and was recently confirmed by satellites as valid over the entire liquid surface of the globe. Evidence that the rise has been happening for much longer than a century has been found along the eastern shore of the North Sea during the last four or five centuries. Note that at any one coastal point, a rise or drop in the shoreline may not be the consequence only of a global sea-level change because shorelines can also respond to upward or downward motions of tectonic plates; the coast of northern Chile, for example, exhibits rapid shifting. Long-term change in ocean levels, however, are not caused by tectonic lifting.

18

GLOBAL WARMING: THE EVERYMAN MISCONCEPTION

*The only ethical principle which has made science possible
is that the truth shall be told all the time. If we do not
penalize false statements made in error, we open the way
for false statements made by intention.*

DOROTHY SAYERS, *GAUDY NIGHT*

THE FIRST WORLD POLLUTES AT A RATE disproportionate to its population, and here we look at measures to change this.

In their recent book, *Betrayal of Science and Reason: How Anti-Environmental Rhetoric Threatens Our Future* (Island Press, 1998), Paul and Anne Ehrlich point out the most mistaken belief—in the United States at least—about conservation, where it is, and where it is going. The Ehrlichs note that "while on the one hand, we applaud the grassroots efforts on behalf of environmental protection (such as curbside recycling, ecotourism, and enthusiasm for things "organic"), we can't help but fear that *these useful but utterly insufficient steps may also help to distract attention from much more basic issues.* Society needs to recognize that to be sustainable, the economy must operate in harmony with rules set by Earth's ecosystems—and needs to act accordingly" (our italics).

In discussing the excesses in waste and pollution primarily found in the first world, we have observed a fundamental difference between areas making considerable progress and those making no progress. During and since World War II, people have made sacrifices for the common good, but they fall into one of three types of sacrifices: The first type we call the province of "Everyman," the sacrifices each of us makes in a spirit of togetherness. They include the recycling of waste products, paying taxes, submitting to automobile inspections and other methods of reducing emissions, and taking individual responsibility for the maintenance of streets, sidewalks, and the collective right of way. They bear a similarity in nature and spirit to the metal and paper drives common during World War II.

But the United States has made little or no progress in two more wasteful areas, and they are the ones in which we waste on a grand scale, and in which we must concentrate our actions. The second area involves the lobbyists who are active and successful in blocking legislation to reduce air, water, and light pollution and fossil-fuel waste. The third area involves collective action in questioning unjustifiable local ordinances.

With respect to the first group—Everyman—we have made great progress in curbing wastes from fossil fuels, although not nearly enough to bring the United States in line with the rest of the world. Americans willing to impose strict laws requiring recycling, for example, are unwilling to give up their lawn mowers, snowmobiles, motorboats, recreational vehicles, SUVs, and vans—all fuel wasters. Most of these machines burn more fuel per mile or per hour than do automobiles. Tax increases could be imposed to discourage their use; indeed, small, fuel-efficient automobiles are now as highly taxed as are the vans and recreational vehicles that consume more fuel per mile than the gas-guzzlers of a generation ago. One long-proposed plan calls for each vehicle with an internal combustion engine to be taxed not only according to its initial cost but also according to its fuel-consumption and emissions rates.

Ski resorts consume vast amounts of energy in providing artificial snow for their customers. It takes about 33,000 kilowatts of electricity to create a foot of snow covering a single mile-long downhill slope; that is more energy than a three-bedroom all-electric home consumes in a year at the same location. If a resort manufactures snow twenty times a season and has twenty trails to cover, it burns more energy in one winter than a small town burns in a year. The skiing industry blossomed when electricity was relatively cheap, and some resorts still have low rates. A willingness to depend on natural snowfall is as justifiable as a willingness to install afterburners on our cars. But boating, skiing, and other such sports are the realm of a minority. Because they are outside the province of Everyman, their participants resist controls.

Heat is the main energy consumer in domestic life. Do homes in the United States really have to be as warm as we keep them during the winter?

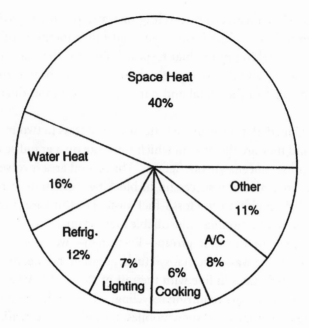

FIGURE 18.1　Home energy use, showing the very large percentage heating uses.

Shouldn't we use the fireplace when we have scrap wood available? We could also improve insulation and keep unused rooms at a lower temperature.

But what about the overall temperatures of public buildings and offices? Heating a building one degree in temperature is several times more expensive than cooling it by the same amount. Yet heat is considered a necessity and air conditioning a luxury. It is difficult to understand this concept in view of the hundreds who die during severe heat waves in even a single city.

In Chicago, for example, 465 people perished in one heat wave in July 1995, half of whom were neither elderly nor ill. According to the Cook County Medical Examiner's Office, more than 50 percent of these victims would have lived had their homes been air conditioned. These fatalities exceeded those of any other single disaster of any kind in the United States since an explosion in Texas City, Texas, in 1947. A 1911 summer heat wave killed more New Englanders than any previous weather disaster there, far more than the much better-known blizzard of 1888 and hurricane of 1938. Yet many heating-system workers raise the thermostat at the request of just one person, despite any number of those who prefer cooler conditions. The average temperature for American homes and public buildings has risen from 68°F (20°C), the definition of "room temperature," to about 78°F (26°C), summer and winter, over the last half of the twentieth century. Despite the growth of the sunbelt, where air conditioning is probably as important as heating, this preference for hothouse conditions wastes an enormous amount of fossil fuel in one form or another.

Most of us are unaware that heating costs far more than cooling (as illustrated in Figures 18.1 and 18.2), but we should note that at the onset of a heat wave, a dwelling warms to the temperature maximum for the day within hours. When cooler weather arrives, that same dwelling takes well over a day to cool, unless there is a brisk wind. Furthermore, electrical appliances that convert electricity into heat are the most expensive to operate: Clothes dryers, toasters, irons, and ranges consume far more energy than washing machines, vacuum cleaners, and television sets.

And finally, we are collectively at fault for failing to ensure that we have responsible ordinances. Many of us, though thrifty as individuals, support wasteful ways of regulating the common good. As astronomers, the authors are aware of the rising light pollution in urban and rural regions in the first and third worlds. Outdoor lighting that shines above the plane of the horizon into the sky—where it is never needed—is a major waste of money and energy. In the United States alone, unnecessary upward-shining urban lighting squanders some 2 billion dollars in fossil fuel energy every year. Full-cutoff shielding lowers energy consumption by a third and darkens sky brightness by some 75 percent. Excessive lighting on bridges and towers disorients migratory birds; as a result, record numbers become exhausted or collide with the structures and die. Lighting planners for one span in Los Angeles called for lights that can be seen with the naked eye from the Moon! And we call ourselves environmentalists?

Manufacturers in North America and in Europe have made great progress in making such items as refrigerators, freezers, air conditioners, coffee machines, and vacuum cleaners more energy efficient. Light bulbs that use half as much current but put out just as much light as conventional incandescent bulbs are also widely available. Why can't we extend these savings to outdoor lighting? Again, the answer lies in the perception that as individuals we can help reduce energy consumption, but when we turn to the collective, the domain of most outdoor lighting, we do nothing. Properly designed outdoor street and private light fixtures have been available for years; let us use them.

The world runs on perceptions even when they conflict with reality. We believe what we want to believe, even when proof beyond a reasonable

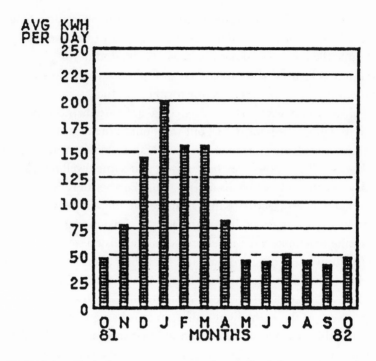

FIGURE 18.2.   Typical energy use over each month of the year in an all-electric home. About half of all energy goes into heating waste.

doubt contradicts our beliefs, a trait we labeled *mumpsimus* in Chapter 3. As to excess outdoor lighting, we are finally seeing a slow but steady change in the belief that more light is better than less light; many state and local governments are passing laws requiring fixtures that do not spill more light than necessary. Utility companies have joined with astronomers, environmentalists, and lighting engineers to raise awareness that excessive outdoor lighting is wasteful and does not deter crime.

Intermodal transport, defined here as intercity freight, is an untouched area of great promise. Freight shipped by rail or on waterways could reduce costs and energy usage for freight transport by about 80 percent. If cargoes were transferred via onloading and offloading from local carriers onto railroads or barges for long-distance intercity transport, additional labor costs would be much more than offset by savings in fuel and money. Furthermore, fewer major road repairs would be required. Two limited-access highways in Connecticut provide an illustration: The Connecticut Turnpike, most of which is also known as Interstate 95, costs many times the upkeep of the Merritt Parkway, which prohibits trucks. These costs are far in excess of the license plate fees collected from heavy vehicles; but by shifting freight to railways, which maintain their own right of way, highway repair and maintenance costs could be slashed. Figure 18.3 is one example of the problem: It compares cars to light trucks, but does not show the low fuel efficiency of heavy trucks.

We should restore the many thousands of miles of railroad track that were abandoned or dismantled in the last half of the twentieth century. Railroads were once the backbone of long-distance cargo transportation in North America and Europe, and they should be so again. Automobiles can be transported on trains, a system becoming more and more popular in Europe and, to some extent, in North America as well. Many long-distance passenger trains include freight cars specially prepared for the fast loading and unloading of private automobiles.

Another problem, this one relating to the Everyman premise, concerns the rise in large fuel-inefficient vehicles. As William K. Stevens points out in his recent book, *The Change in the Weather* (Delacorte Press, New York, 1999), the moderately successful reduction by the Clinton-Gore administration to roll back greenhouse gas emissions to levels near those of 1990 (as is called for under the Kyoto accords) has been more than offset by the energy demand imposed by the resurgence of gasoline-wasting vans, sport utility vehicles, minivans, and pickup trucks. These vehicles now account for over half the sales of passenger cars, but they are subjected to federal fuel-efficiency standards less stringent than those for smaller passenger cars. Environmentalists unaccountably tolerate these gas-guzzlers, and even desire to own them, while willingly supporting emissions-control measures for cars in the main.

In several large cities, for instance Santiago de Chile and Caracas, drivers must submit to a rotation system governed by an odd-even system in the last digits of license plate numbers, although the effectiveness of this measure is doubtful. First, the system can work only if alternative methods of transportation are available; second, affluent drivers dodge the system and buy second cars. Good public transport is a much better alternative for metropolitan areas; the metro or subway is an ideal solution for cities and good bus service benefits smaller towns. Fast lanes, already widely used in the United States, are restricted to vehicles with at least two occupants, but this system works best with car pools. And because fast lanes demand individual initiative, they are not a major source of conservation. As we mentioned in Chapter 16, when Connecticut has removed its tollgates, the

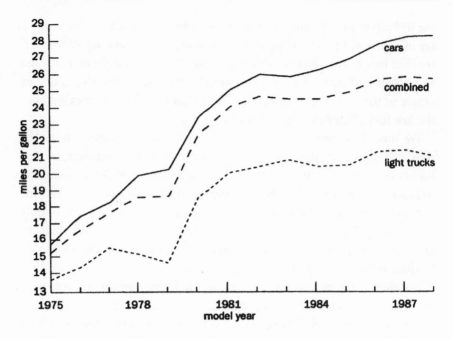

FIGURE 18.3   Trend in fuel efficiency for cars and light trucks. Similar data for heavier trucks are not available.

immediate results were reduced air pollution and accident rates, a much more useful contribution to energy savings.

•   •   •

Two hard winters in eastern North America brought record snowfall and near-record cold temperatures to New England and many other places. During the winter of 1993–1994, Hartford, Connecticut, recorded the coldest two-month average of the twentieth century, with a record 85 inches of snow. Two years later (1995–1996), a whopping 115 inches of snow fell, 30 inches more than fell in 1993–1994, and tripled the average amount.

What about these winters? Shouldn't they be warmer and milder than their predecessors? Can global warming be real or does this disprove it? We know that further global warming must begin to affect winter weather as it already has increased summer heat. We have already explained that most global-warming models predict that temperatures will increase more rapidly in the cold polar regions than at midlatitudes, and more there than in the Tropics. But several winters in the 1990s were severe. Supporting evidence in the form of frost damage reaching into southerly counties in the orange-growing areas of Florida, for example, suggests colder winters as well as hotter summers in our recent past. Today, Florida's orange-growing belt lies two counties to the south of its location in 1950.

The media have often doubted the evidence for global warming based on these long, cold winters. But colder winters are not less consistent with the effects of greenhouse gases than very hot summers for two related reasons. The first is that overall climatic trends can be masked by short-term effects that temporarily overwhelm them. It may take longer for the warming trend to affect winter weather, whereas it is already evident in the summertime. And, as Stevens states, "the behavior of snow in a warming climate is paradoxical." Very cold air cannot hold enough moisture to sustain major snowstorms; with warmer conditions, larger snowfalls are more likely—contrary to the public's perception.

Global warming has not and will not soon cancel wintry conditions. Snowstorms of greater severity can be products of warmer weather.

It is helpful to examine the nature of winter weather systems, especially in the midlatitudes. In summer, weather systems are comparatively large and slow-moving. Endless hot, hazy days pass with only an occasional shower, and rarely a front of any consequence. Few days in summer are fully overcast and gray from dawn to dusk; most are hazy, the Sun visible, if not bright. Air masses are huge and the fronts between them are rare and ineffective. The best-known and most prominent summer feature affecting the United States is the infamous Bermuda High, which pumps its hot, sticky air over much of North America, frequently for weeks at a time. All summer long, maritime tropical air dominates and is only occasionally invaded by an outbreak of cooler, drier continental polar air. Summer rain falls mostly as brief showers and thunderstorms.

In winter, cold continental polar air pushes south again and again into the northern states and sometimes beyond them. Weather systems are smaller and more violent, and they move much faster; it is common for two storms to pass a given area in a week. Frequent cold and warm fronts can sometimes usher in severe weather. The storms, the nor'easters we discussed in Chapter 1, can dump rain or snow in great amounts for days at a time.

In the winter of 1995–1996, the northern states were buried in snow, primarily because a low-pressure system formed high in the atmosphere and sat for months over eastern North America. This polar vortex allowed cold air to pump down from Canada and invade the East and Midwest, where it met the milder Atlantic air; the boundary between the two air masses became a track for one storm after another.

This extreme weather may happen again in the winter of 2000–2001, or it may not happen for several years. Next time we experience a long stretch of harsh winter weather, we should recall, and weather forecasters should remind us, that local and temporary winter conditions replace the pervasive warm-up that dominates the long, hot summer. Such regional irregularities are no proof that global warming has ceased. Recall that the climatic conditions over a single region and over a modest time interval are not representative of the entire globe.

Scientists already know the result of ignoring the data, qualitatively, if not quantitatively. We know that the twenty-first century will consummate the "sixth great mass extinction," in Richard Leakey's words, of the world's species since life outgrew its one-cell forms a half-billion years ago, possibly outperforming even the deadly comet/asteroid calamity of 65 million years ago. All the remaining great land mammals will soon be gone; the elephant, the giraffe, the hippopotamus, and the rhinoceros are about finished already. Beaches will be covered with jellyfish and crude oil, and the grapes of wrath will return to Oklahoma.

Heat is coming, and it will beat down relentlessly and with the same intensity that drove Albert Camus's stranger, Meursault (in *L'Étranger*) to commit a senseless murder on an Algerian beach; the kind of heat in Tennessee Williams's *Suddenly Last Summer* that drove febrile young men to storm a hill in chase of Sebastian, beating their rattles and noisemakers, until they caught and cannibalized him. Only the extreme cold, stormy, frostbitten conditions that met Rob Hall and Scott Frazer on Mount Everest, or the waste of ice known to Robert F. Scott and his doomed men in Antarctica, can compare with heat in the lexicon of evil in nature as well as art and literature through the ages. Hell is with reason portrayed as hot.

Let us peer at our future through Meursault's tortured eyes: "The Sun was starting to burn my cheeks, and I could feel drops of sweat gathering in my eyebrows. The Sun was the same as it had been the day I'd buried Maman, and like then, my forehead was especially hurting me, all the veins in it throbbing under the skin. It was this burning, which I couldn't stand anymore, that made me move forward. I knew that it was stupid, that I wouldn't get the Sun off me by stepping forward. But I took a step, one step forward." The murder that ensued was as much as anything a product of this merciless heat. During the Chicago heat wave of 1995, mentioned earlier in this chapter, even a rapid and timely trip to the emergency room did not save those stricken by summer heat. Nearly half the victims died within a year, and most survivors were permanently disabled. Heat is a greater killer than any other weather phenomenon.

Those who would tag global warming as a scheme of "environmental wackos (who) . . . want to attack our way of life," as one of our commentators would have it, are performing a great disservice to themselves and their societies. They are part of the problem, not the solution. Scientists of many persuasions, nationalities, and ideologies around the world have spent years of their lives documenting the rise in global temperatures and correlating them to human activity. Their findings, although incomplete, are convincing to the point that we must set aside preconceptions about departing from "business as usual." Business as usual is itself a deliberate action and may not be an option for long.

How many beliefs with far less credibility and factual support than global warming do we hold dear? In the United States, for instance, the belief that the death penalty is something more than vengeance; the laws, now mostly repealed, to lower all speed limits to 55 miles (89 kilometers) per hour in order to save lives and fuel; and the belief that life and consciousness begins at the moment of conception. All these have far less scientific support than the premise that temperatures and ocean levels will rise to dangerous and unacceptable levels in the twenty-first century. None of the shibboleths of our day, or the medieval approach to science in which every proof must be fixed before a theory departing from dogma can be accepted, are fit for our consideration. Science has confirmed and warned that we face this problem too soon to ignore it any longer.

As Vice President Albert Gore mentions in his book, *Earth in the Balance*, one of his college professors, Roger Revelle, "convinced the world scientific community to include as part of the International Geophysical Year (1957–58) his plan for regularly sampling carbon dioxide concentrations in the atmosphere." At that time, his colleague, C. D. Keeling, began the series of now-famous measurements atop Mauna Loa in Hawaii. The graph of those ever-escalating concentrations (shown in Figure 1.1) has become a symbol of the disaster awaiting us should we opt for business as usual.

Not all published works see the increase of carbon dioxide and global warming as harmful or of concern. In his *Climate of Fear: Why We Shouldn't Worry About Global Warming* (Cato Institute, 1998), Thomas Gale Moore, a senior fellow at the Hoover Institution, opposes action to delimit further greenhouse gases in the air. In the first chapters, he discounts the prediction that a global warm-up is underway; but later on, in a partial antithesis, he judges a warm-up beneficial. He then appears to refute his case for beneficial warming through his thorough discussion of the economic displacements such a warm-up might cause. We are not the simple agrarian societies of 5000–3000 B.C., when warm weather fostered and nurtured the first river-valley civilizations.

Moore's repeated appeals to the emotions are troubling: "Vice President Gore has divined"; "A media chorus has fanned the fear"; and "The (Kyoto) conference had degenerated into a mix of revival meeting and guerrilla warfare." These and a number of other comments have no place in an academic treatise. Although extremists on all sides of the issue employ grandiloquent discourse, global warming should not become a partisan issue, nor should big business support or cite scholarship without critically evaluating it. Carl Sagan's statement that "extraordinary claims require extraordinary evidence" should apply to all aspects of the global warming question. If it is too early to make a case for warming beyond any reasonable doubt, we should nevertheless continue the investigation and take action at some point. The higher the oceans rise and the more severe droughts and hurricanes become, the harder it will be to restore the old climate.

A FINAL STATEMENT

The United States is one of more than 160 nations participating in international conferences (such as Kyoto) dealing with climate change, but we cannot adopt any of the accords without approval from the Senate. Significant international developments include British Petroleum's commendable support for reducing fossil fuel usage; we hope that other large oil concerns will soon follow BP's lead.

We believe that Congress's refusal to bring the accords up for discussion is indefensible. Much of the culpability for the severity of more disasters such as hurricanes Georges and Mitch and the fires of Southeast Asia will be laid squarely at our door if we keep stalling.

Certainly, the case for global warming is commanding. Can the parallel increase of the average temperatures around the globe and the carbon dioxide content in the atmosphere, after a long period of stability, be coincidental? Or does the one give rise to the other, as theories predict and observations of this and other planets bear out? If the latter is the more probable explanation, we are making an unplanned and global historic experiment. If current emissions trends continue, the gain in heat energy from greenhouse gases may double by the year 2030; in that event, the average temperature will rise by another 1°F to 3°F around the world by that time. Polar ice caps will melt further and ocean levels will rise and inundate low-lying coastal areas and some island republics; prolonged droughts and more tropical storms will accompany these and other changes in climate.

The amount of global warming is not fully derived; few major scientific paradigms are clearly known. The heliocentric theory of Copernicus had to wait two to three centuries for firm observational authentication—long known to exist but then undetectable—but its widespread acceptance came about in less than a century. Delay made here in the belief that some final vindication must first be realized will not fly. Irving M. Mintzer, editor of a survey about the many aspects of global warming, *Confronting Climate Change* (Cambridge University Press, 1992), states that "human activities are changing the composition of our atmosphere at an unprecedented rate. If current trends continue, our planet could face a climatic shock unlike anything experienced in the last 120,000 years. It would not be felt as an immediate blow; that is, a shift from status quo to catastrophe. The climate cannot disappear like an endangered species. Nor can it explode like a runaway reactor. Nevertheless, the risks of climate change—rapid by geologic and climatic standards—are rising rapidly in our time. In this context, 'shock' is an appropriate term." This is a problem without precedent, unique in its potential for the devastation of the planet. Our species, all 6 billion of us, is involved together as never before.

APPENDIX I:
MASS AND WEIGHT

Weight is just a measure of the mass, the amount of matter, in a body as measured at the Earth's surface. On the Moon, one's mass would remain the same as on the Earth, but the weight would be less because the Moon is smaller and less massive and has a smaller gravitational force as a result. At its surface, the Moon attracts with only one-sixth the pull of the larger Earth; a 150-pound person would weigh only 25 pounds there. That is why astronauts can wear spacesuits weighing more than they do and yet function easily on the lunar surface.

Air pressure decreases with altitude, and it does so in a predictable and calculable way. The top of the atmosphere is indefinable: No boundary can be set between very thin air and empty space; one merges into the other. An arbitrary altitude of about 200 kilometers (124 miles) might do as an approximation because air affects artificial satellites below this limit by friction, dragging them slowly downward to eventual impact with the surface.

The amount of atmosphere from bottom to top is well known. If all air were compressed to its density at sea level, that is, if it were of constant sea-level density to its top, with only empty space above, it would extend only to 5 miles (8 kilometers) above the ground. This constant-density height is called the scale height; this and the decline in air density with elevation are both shown in Figure 2.2.

For more than a century, we have been aware that air pressure at any one spot varies slightly over time. Throughout a day, a week, or a year, it changes by as much as 2 to 3 percent of its average and we get a succession of high and low pressure regions (highs and lows) as a result. The changes can be detected by noting the corresponding variations in a 34-foot water column, which weighs just as much as a column of air of the same cross-section extending to the top of the atmosphere. But a column 34 feet high (just over 10 meters) is not a practical thing to construct or observe. If a tube of this height open only at the bottom were set into a tub filled with water, we would find that the air pressure on the water surface would force the water to rise up into the tube to 34 feet.

A tube with a much denser liquid would be far more practical and easier to handle. The densest common liquid is mercury, 13.7 times the density of water. We need a column of mercury only about 1/13.7 of 34 feet, or 30 inches (76 centimeters) tall to record changes in air pressure. Such instruments, called barometers, are available, but even a 30-inch tube is bulky and expensive. Smaller, cheaper barometers are more widely available; these are known as *aneroid barometers* and consist of a hollow, airtight cylinder with a spring that registers changes on a dial. The aneroid barometer is calibrated in the same units (i.e., a 30-inch or 76-centimeter scale with normal variations) just as if it were a tube of mercury.

All this leads to the many ways we can measure air pressure. Pressure is defined as the force per unit area exerted by the air. The average pressure of air at sea level amounts to a pressure or weight of 14.7 pounds per square inch, or 1.034 kilograms per square centimeter. But pressure is just as commonly expressed another way. Mercury barometers are so commonly used that air pressure is also given as 30 (strictly 29.92) inches (76 centimeters) of mercury because it balances a mercury column of that height. Meteorologists most frequently use the millibar (one thousandth of a bar). The average sea-level pressure is 1013.2 millibars. Other units are sometimes used; the entire bewildering mess of units has resulted from considering the impact of air in terms of pressure, weight, or force. Each comes with at least one set of units in the metric system and another in the British system.

A specific mass of air can be described by three variable quantities: the volume it occupies, V, its temperature, T, and its pressure, P. If air is heated in an enclosed volume, the pressure rises right along with the temperature. If the volume is increased, as it might be in the atmosphere with no limits in space, the temperature drops and the pressure falls along with it. Thus these three variables are closely interrelated by the well-known gas laws, such that if two are given, the third can be calculated. In simplest form, the gas law states that $PV = RT$, where R is a constant factor.

The air pressure at any distance above sea level is determined almost entirely by its altitude, with a maximum variation over time of only a few percent. Aside from these small variations, the pressure falls off in a predictable manner. It is hard to name another factor that plays so important a role in any understanding of the vertical structure of the atmosphere. The variation of the air density with elevation is close to that of the pressure with elevation, but it is not identical to it. Some 80 percent of the atmosphere lies within about 8 miles (12 kilometers) of the surface, even though it extends upward to more than twenty times that elevation. Because such features as clouds, rain, and snow require relatively dense air for their formation, they are confined to these low, dense layers of the atmosphere known as the troposphere.

APPENDIX II:
AIR PRESSURE AND
DENSITY

Pressure and density are related to each other and to the temperature; they are all interdependent, as the ideal-gas law makes clear. Let us consider a small parcel of air anywhere in the atmosphere with its physical parameters (temperature, pressure, and density) identical to those of the air surrounding it. If for any reason our parcel should rise just a bit, it will find itself surrounded by air of a slightly lower pressure. Thus the parcel will expand, seeking to equalize its pressure with the neighboring air. In doing so, and as a consequence of the gas law, it will also lower its temperature as well as its density because it now occupies a somewhat larger volume. At this point, three possibilities may occur:

1. The parcel finds itself heavier than its new surroundings and will sink back to its previous level. The stratification of the air is stable; that is, little vertical convection will occur. One such condition occurs whenever the temperature stratification is inverted, namely, whenever the temperature increases with altitude. Temperature inversions are frequently found at nighttime over the desert floor, although they can form almost anywhere in the atmosphere. This is the condition known as *stable equilibrium,* and it is analogous to a cone standing on its base. If the cone is tipped or disturbed, it returns to its initial position immediately.

2. The parcel's density remains equal to that of the surrounding air; it will neither continue to rise nor will it sink back to its original elevation. Such a stratification marking the border between stable and unstable conditions is called an *adiabatic distribution.* Most of the time, the Earth's atmosphere has a nearly adiabatic structure but never quite reaches it because of the many variable influences upon it. This state is one of

neutral equilibrium and is similar to a cone lying on its side—it rolls to a stop rather quickly, but does not return to its original position.

3. The parcel finds itself lighter than its surroundings and thus continues to rise. If conditions are favorable for its rise, and frequently they are, the parcel may be on a long skyward journey that ends when it finally hits a temperature inversion. Such convection currents form the principal vertical stirring mechanism in the Earth's atmosphere. Among the most striking manifestations of convective currents are cumulus clouds, the fleecy puffed clouds that are so common during a sunny summer afternoon. Here is unstable equilibrium, at least until stability is restored, much as a cone balanced on its point or vertex. A slight push sends it toppling until it comes to rest in the neutral position.

Vertical motion would be initiated in an adiabatically stratified atmosphere by any form of energy input. Solar radiation and reradiation from the Earth's surface are the overwhelmingly largest sources of energy input into the atmosphere; the amount of heat flux caused by residual heat and radioactive processes coming initially from the interior of the Earth is negligible.

Solar radiation received at the top of the atmosphere is highly variable from day to night and through the yearly seasons, with long-period variations caused by changes in the orbital configuration of the Earth and slight physical variations in the energy output of the Sun itself.

The energy flux that reaches the top of the atmosphere from the Sun has a certain spectral distribution over the different colors (curve at left in Figure 3.2) that is primarily a result of the Sun's temperature. This curve varies slightly with the eleven-year cycle of solar activity. These changes occur mostly in the ultraviolet region where they may amount to as much as 20 percent of the normal radiation in that part of the spectrum. Across the entire spectrum, variations would not even make up 0.1 percent of the total radiation; that is, the distribution of intensity over the different colors or wavelengths.

Once a beam of radiation enters the Earth's atmosphere, its future depends on four parameters:

1. the angle of incidence;
2. the presence or absence of clouds;
3. the absorption and scattering properties of the air; and
4. the absorption, reflection, and scattering properties of the surface.

The path length of a beam of radiation traversing the atmosphere in the absence of clouds reaches a minimum for vertical incidence, which occurs when the Sun is at the zenith, the point directly overhead. It becomes longest when the Sun is near the horizon. Naturally, the absorption or loss of radiation increases with the path length. For example, when the Sun is at 60 degrees from the zenith (i.e., 30 degrees above the horizon), the loss is exactly twice the loss at the zenith.

The atmospheric absorption, usually referred to as *atmospheric extinction,* depends strongly on the radiation's wavelength, a dependence closely related to the chemical composition of the air. As mentioned above, dry air is composed almost entirely of only two gases, diatomic nitrogen (N_2) and diatomic oxygen (O_2), with a small amount of the inert gas argon (Ar). If these gases were the only ones present, the spectral distribution of the radiation reaching the ground in the case of vertical incidence would be given by the solid curve just below the dashed curve in Figure A.1; the space between that solid curve and the dashed curve represents the energy absorbed by the air. The real spectral distribution of the radiation reaching the ground—again for the case of vertical incidence—looks quite different, as shown by the lower dashed curve in Figure A.1. (See page 210.) The indentations in this curve, representing the additional absorption at specific wavelengths, are mostly caused by either water vapor, which can be quite abundant at times, or by carbon dioxide (CO_2), ozone (O_3), methane (CH_4), or nitric oxide (N_2O). Only traces of these gases are found in the atmosphere, but their absorbing capacity is so enormous that they retain about as much radiation as do the much more abundant N_2- and O_2-molecules.

After having traversed the entire atmosphere and experiencing the losses described above, a beam of light reaches the land or water at the surface. If the ground is covered with snow or ice, much of the radiation will be reflected upward, where it passes through the atmosphere once more; it undergoes further loss, caused by absorption, before leaving the Earth altogether.

At this point another important process comes into the picture. As mentioned above, the ground also radiates most of its emission in the far infrared. Although the atmosphere is fairly transparent for visible light, which comprises most of the solar radiation, this is not the case for infrared light, at which the air is practically totally opaque; the opacity is mostly caused by those trace gases that produce the extra absorption of the incoming radiation. The total energy taken out of the outgoing ground radiation exceeds that taken out of the incoming solar radiation to such an extent that the former is the primary mechanism responsible for heating the air. This process is known as the greenhouse effect and the gases responsible for it as the greenhouse gases.

APPENDIX III:
ABSORPTION AND
CONDENSATION

The abundance of the greenhouse gases varies only slowly with time over the years, with the exception of water vapor, the concentration of which can change rapidly with time and location, both horizontally and vertically. Water can be present in the air in any of its three physical forms: liquid water, ice, or vapor. Liquid and solid particles can be so small that they float in the air, thus forming part of it in the form of clouds, fog, or haze.

A given air mass, let us say a cubic meter, can only retain a certain maximum amount of water in the form of vapor. This amount increases in a characteristic and predictable way with the temperature. From this, we can readily deduce one of the fundamental processes going on in the air, namely, the cloud formation. Let us imagine a parcel of air with a certain amount of vapor, well below the maximum allowance. Let this parcel rise a little for whatever reason. If the stratification of the air is appropriate, the parcel may then move upward a long way, as explained above. On its way up, it continuously lowers its pressure and temperature, and with it its maximum vapor allowance. Eventually, it reaches an elevation and corresponding temperature such that its vapor content equals its maximum allowance: It has reached *saturation.* Any further rise leads to an excess of water vapor and condensation, or the formation of microscopic water droplets. Because this condition starts at a well-defined elevation, clouds are often flat bottomed, thus marking that elevation. Supersaturation alone, however, does not cause condensation; *condensation nuclei* help form the initial droplets. Once these are present, other processes help make them grow effectively. The original nuclei may be dust particles picked up by the winds in the desert, or they may be salt crystals ripped from the crest of a wave during a storm over the ocean. They may be of extraterrestrial origin: debris left by a comet or meteor; they may be of volcanic origin, or they may be manmade. Usually the particles, called *aerosols* whatever the source, are a combination from several of these sources. We will discuss the manmade aerosols later. If no nuclei are available—a rare condition nowadays—no condensation occurs and the air may become highly super-saturated. If nuclei are overabundant, however—a normal condition now caused by the large human input—and if these nuclei are acids, condensation may occur well before saturation is reached.

Clouds have an important impact on the heat balance of the air as well as of the ground. Most of the solar radiation intercepted by clouds is reflected back into space and does not contribute to the heating of the air below them or to that of the ground. They do contribute to the greenhouse effect, however, by reflecting radiation from the ground back to the ground. At the same time, they emit radiation—again in the infrared—and to a lesser degree, such reflection is true of the air itself.

Absorption of radiation emitted by the ground is certainly the most efficient and far-reaching energy input mechanism into the air, but it is not the only one. Transfer of energy by contact with the ground can, under certain circumstances, become very important. Such is the case when there is a large temperature difference between the air and the ground, either positive or negative. When the ground is very cool, a strong temperature inversion forms just above it; the inversion may be stable enough to last for a long time. An overheated ground leads to turbulent upward motions and the apparent blurring of distant and, sometimes, nearby objects. Other atmospheric optical phenomena such as a *Fata Morgana*, or reflections off the surface of a highway, are also related to anomalous stratification near the ground caused by direct energy transfer.

Evaporation of water from the surface, either from a body of water or ground moisture, is another energy transfer mechanism we have to take into account. Evaporation is really a process that injects new molecules of water into the air, and these particles carry a temperature message with them; they represent a temperature higher than that of the surface they came from, thus cooling the latter at the same time. We feel this surface-cooling effect when we expose wet skin to the wind.

The energy exchange mechanisms occurring at the surface hint at the possibility of a coupling between the troposphere and the oceans just beneath. The existence of such a coupling has been known for some time, and terms such as *maritime climate* and *continental climate* point out where this coupling works. There is even more to it than this. If we calculate the expected yearly average temperature at a given geographical latitude only according to the total amount of solar radiation received during a year, we find that the equatorial zones are much cooler than they should be, while the polar regions are far too warm. This means that energy is effectively transported from the Tropics to the poles. In principle, both the atmosphere and the ocean currents can transport energy. Measurements show that both are involved, the ocean currents carrying the major portion.

The mechanisms of energy input discussed so far apply in their strictest sense only to the lowest 9 to 16 kilometers (6 to 10 miles) of the atmosphere, the troposphere. The stratosphere, the layer lying just above the tropopause, receives heat input from above, directly from the Sun. Thus a gigantic temperature inversion forms, which extends to an elevation of about 50 kilometers (31 miles). This layer of the atmosphere is called the stratosphere because a temperature inversion constitutes a stable stratification. The major portion of solar ultraviolet radiation is absorbed within the stratosphere, and this absorption is caused only by one type of molecule, ozone (O_3). Ozone is oxygen in triatomic, not the usual diatomic, form. Ozone is responsible for most of the heating in the stratosphere because it is such an efficient absorber of the ultraviolet radiation. As stated earlier, the ultraviolet radiation of the Sun varies considerably with the solar cycle and, naturally, this affects the heat balance of the stratosphere. Only recently has it been discovered that periodic depletion of the stratospheric ozone is occurring every year over the Antarctic. Scientists are now certain that this depletion is caused by input from human sources. The situation has created great concern worldwide because of the harmful effects of the ultraviolet radiation that might leak through the stratosphere and reach the ground. We have dedicated an entire chapter of this book to the ozone problem.

The top of the stratosphere is marked by a temperature maximum at around 50,000 meters (164,000 feet). In the layers farther aloft (the mesosphere, the thermosphere, and the ionosphere at the very outer limit of the atmosphere) the air is so rarefied that the gas laws we have used for the troposphere no longer apply. In addition, although the stratosphere is known to be strongly coupled to the troposphere and thus participates in weather-making, the uppermost layers of the Earth's atmosphere are of marginal influence in the weather below.

As we went along explaining the most important physical processes that occur in the Earth's atmosphere, we found several that may be altered or even caused by human activity. In particular, mankind is producing a variety of greenhouse gases. It is obviously important to know whether the manmade greenhouse gases make a significant contribution to those already existing. We have pointed out, too, the significant manmade contribution to the atmospheric content of condensation nuclei. Although the first produces a global warming, the latter increases cloudiness, thereby causing cooling. It is now well known that both effects are well above the significance level and are increasing at an impressive rate. What their combined effect has brought about already and what it will be in the near and far future is no longer a subject of speculation; it is the subject of intensive research in more than one direction.

Each of the manmade injections into the atmosphere of greenhouse gases and aerosols generate a major impact on the climate of the planet. The two effects act in opposite ways, an unusual circumstance, and almost compensate each other, leading to a nearly undetectable total effect. Should either of the two be greatly reduced or eliminated, therefore, the other would predominate and modify the climate in one direction or the other. We cannot continue to mess up the atmosphere forever. Warming and cooling are not the only impacts of human influence on the weather: They affect our health and that of all other life. Thus we must find a careful balance and reduce, step by step, our influence on the atmosphere, which is shared by all forms of life.

APPENDIX IV: THE CORIOLIS EFFECT AND THE PREVAILING WESTERLIES

As we explained earlier, heating or cooling of an air mass by whatever process will bring about a change in its pressure. The heating or cooling is not uniform over the entire globe, but often can show large local variations both in space and time; these are caused by the differences in cloudiness and albedo. Significant pressure differences can build over short distances. These differences set air into motion; that is, they cause wind, which tends to equalize pressure differences.

The principal pressure gradient spanning the Earth in each hemisphere is caused by summer-winter temperature differences in the midlatitude regions. It is always hot in the Tropics and cold in the polar regions, but somewhere between the two, the difference can be acute. This temperature gradient sets up a pressure gradient, and with it, a pressure-gradient wind that seeks to blow poleward in each hemisphere. The pressure-gradient wind is strongest wherever the difference is greatest, and it moves northward or southward with the seasons. Winter in the Northern Hemisphere finds this gradient strongest between Minnesota and Florida, and between northern Europe and North Africa, for example; in the summertime, the gradient and therefore the wind moves northward into northern Canada and Scandinavia.

But in comes the influence of the spinning Earth. The force driven by it is known as the *Coriolis Effect*. The Coriolis Effect redirects anything moving freely, whether a missile or a wind. Regardless of the direction of the compass, missiles and air alike are diverted to the right in the Northern Hemisphere, and to the left in the Southern Hemisphere. This effect rotates expanding high-pressure anticyclones in a clockwise direction in the Northern Hemisphere while contracting low-pressure cyclones rotate counterclockwise. The sense of rotation reverses in the Southern Hemisphere. Hurricanes, being strong low-pressure storms, follow the same rule, but very small-scale spinning atmospheric systems such as tornadoes, dust devils, or water systems (such as water in a draining bathtub) can spin

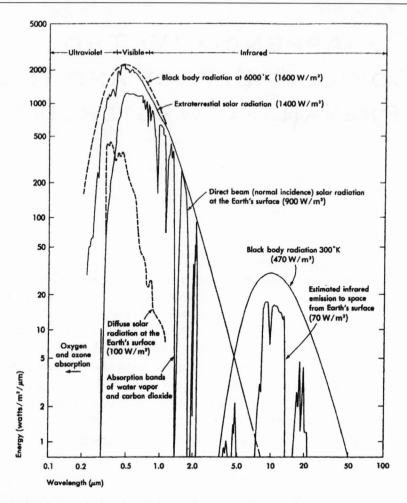

FIGURE A.1   calculated and observed spectra of energy from the Sun and Earth.

in either direction because their cross-section is just too small for this effect to take hold.

The resulting wind, affected as it is by the pressure-gradient wind and the Coriolis Effect, blows generally from west to east in the midlatitudes in both hemispheres and is known as the *geostrophic wind,* or more commonly, the *prevailing westerlies.* Deviations from a straight west-to-east circulation occur frequently, but the overall trend lies in that direction. This is why weather usually comes to residents of the temperate zones in the intermediate latitudes from the west.

APPENDIX V:
FURTHER REMARKS ON
THE CAUSES OF MASS
EXTINCTIONS

A key argument in the controversy will be the answer to the question, How sudden were these extinctions and how fast was the change of the global climate? The geological stratification alone cannot distinguish between events that took only hours to happen and others that were spread over thousands or even millions of years. A major impact will have an immediate and catastrophic effect on weather conditions and could wipe out many species in a very short time. Increased volcanic activity will change the meteorological conditions more slowly and less drastically, but can last for a long time. Isotope abundances estimated for times right across the K/T boundary indicate that the temperature drop occurred rather slowly over thousands of years. Maybe there is some truth in both theories. It could well be that the impact of a celestial body produced the first climate shock as well as the abundance of iridium and microdiamonds and simultaneously triggered tectonic-plate motion. The subsequent volcanic activity would then maintain a contaminated atmosphere for a long time, thus extending the period of climate alteration.

Additional information was recently extracted from a study of microfossils at the K/T boundary in ocean sediments. The relative abundance of a strontium isotope shows a sharp increase right at the boundary, with a subsequent decline lasting for about 10 million years. The strontium isotope forms through radioactive processes. One of these occurs at the surface of the Earth through interaction with the cosmic radiation. Because this radiation does not penetrate into deep layers, the strontium isotope is more abundant at the surface. Weathering action carries surface material into the rivers and from there eventually into the oceans, where it finally settles in the sediments at the bottom, including its isotope-enriched strontium. Thus an increase of the weathering action on the surface can subsequently lead to an increase in the relative isotope abundance in the sediment. The effectiveness of weathering depends upon the temperature fluctuations that break up the surface of rocks and on the rain and wind, which carry the broken material away. It also depends on the acid content of rain. Acids effectively destroy the surface of solid rock. Both impacts and volcanic activity produce large amounts of oxides of nitrogen in the air; the oxide molecules rapidly combine with water to form acid rain.

Heavy acid precipitation at the onset of the event may have brought about a sudden increase in strontium isotope abundance on the floors of the oceans soon thereafter. Even if the acidity of the precipitation ceases after only a few months, however, it will take millions of years before the entire additional eroded material has been removed from its original sites and buried at the bottom of the sea.

APPENDIX VI:
INSTANCES OF
GLOBAL WARMING

The December 13, 1999 issue of *Time* mapped twenty-seven ways in which global warming has already been confirmed. The first fifteen of these are labeled *harbingers*, in the sense that they are becoming more frequent and widespread with time. The last twelve are called *fingerprints*, direct manifestations of global warming. The harbingers are

Vanishing animals:

1. In California, Edith's Checkerspot butterfly has disappeared from lower elevations and southern limits of its range.
2. In Antarctica, Adélie penguin populations have declined by 33 percent in twenty-five years because the sea ice where they live is shrinking.
3. Canadian Arctic caribou numbers dropped from 24,000 in 1961 to 1,100 in 1997, mostly because heavy snowfalls and freezing rain covered their food supply.

Storms and floods:

1. From August 15 to 17, 1998, a storm dumped nearly 1 foot (30 centimeters) of rain on Sydney, Australia, three times as much as normally falls during that entire month.
2. In Korea, severe flooding struck during July and August, 1998. On some days, rainfall exceeded 10 inches (25 centimeters).
3. In February 1998, 21.74 inches (55.22 centimeters) of rain fell on Santa Barbara, California, its highest monthly total on record.

Spreading disease:

1. In 1997, hundreds of people died of malaria in the highlands of Kenya, where the population had not previously been exposed.
2. In the Andes of Colombia, mosquitoes that can carry dengue and yellow fever, once limited to altitudes no higher than 3,300 feet (1,000 meters), appeared at altitudes of 7,200 feet (2,195 meters).
3. In 1997 in Indonesia, malaria was detected for the first time as high as 6,900 feet (2,100 meters) in Irian Jaya province.

Droughts and fires:

1. More than 1.2 million acres (500,000 hectares) of forest burned in Spain in 1994.
2. In 1998 in Mexico, 1.25 million acres (506,000 hectares) went up in flames during a severe drought.
3. Up to 2 million acres (800,000 hectares) of land in Indonesia burned in 1998, including parts of the already devastated rain-forest habitat of the Kalimantan orangutan.

Earlier spring:

1. Average of 8.8 days.
2. During eighty-two years on record, 4 out of the 5 earliest thaws on the Tanana River in Alaska have occurred in the 1990s.
3. The length of time Mirror Lake in New Hampshire is covered with ice has declined about 0.5 day per year during the past thirty years.

Heat waves:

1. In 1998, Lhasa, Tibet had its warmest June on record. Temperatures exceeded 77°F (25°C) for 23 days.
2. In Cairo, 1998 brought the warmest August since records have been kept. Temperatures reached 105.8°F (41°C) on August 6.
3. In 1999, New York City had its warmest and driest July on record, with temperatures climbing above 95°F (35°C) for 11 days.

Rising seas:

1. Saltwater inundation in Bermuda from the intruding ocean is killing coastal mangrove forests.
2. Sea-level rise at Waimea Bay, Hawaii, along with coastal development, has contributed to considerable beach loss over the past ninety years.
3. The shoreline of Fiji has receded 0.5 foot (15 centimeters) per year for ninety years, according to local reports.

Melting glaciers:

1. The Gangotri glacier in India is retreating 98 feet (30 meters) per year.
2. In the Caucasus mountains of Russia, half of all glacial ice has disappeared in the past hundred years.
3. The Qori Kalis glacier in the Andes Mountains of Peru is receding about 100 feet (30.5 meters) per year, a sevenfold increase in rate since the 1960s and 1970s.

Polar warming:

1. In Barrow, Alaska, the average number of snowless days in summer has increased from fewer than 80 in the 1950s to more than 100 in the 1990s.
2. The area covered by sea ice in the Arctic Ocean declined by about 6 percent from 1978 to 1995.
3. In Antarctica, nearly 1,150 square miles (2,980 square kilometers) of the Larson B and Wilkins ice shelves collapsed from March 1998 to March 1999.

BIBLIOGRAPHY

Ahrens, Donald C. *Essentials of Meteorology: An Invitation to the Atmosphere.* Saint Paul: West, 1993.

Alvarez, Walter, *T. Rex and the Crater of Doom.* Princeton: Princeton University Press, 1997.

Battan, Louis J. 2d ed. *Fundamentals of Meteorology.* Englewood Cliffs, N.J.: Prentice-Hall, 1984.

_____. *Weather in Your Life.* San Francisco: Freeman, 1983.

Brooks, C.E.P. *Climate Through the Ages.* 2d ed. New York: Dover, 1970.

Burroughs, W.J. *Watching the World's Weather.* Cambridge: Cambridge University Press, 1991.

Diamond, Jared. *Guns, Germs, and Steel.* New York: W.W. Norton, 1997.

Ehrlich, Paul R., and Anne H. Ehrlich. *Betrayal of Science and Reason: How Anti-Environmental Rhetoric Threatens our Future.* Washington, D.C.: Island Press, 1996.

Firor, John. *The Changing Atmosphere: A Global Challenge.* New Haven: Yale University Press, 1990.

Gore, Albert. *Earth in the Balance.* New York: Houghton Mifflin, 1992.

Hoyt, Douglas V., and Kenneth H. Schatten. *The Role of the Sun in Climate Change.* Oxford: Oxford University Press, 1997.

Imbrie, John, and Katherine Palmer Imbrie. *Ice Ages: Solving the Mystery.* Cambridge: Harvard University Press, 1979.

Lamb, H.H. *Climate History and the Modern World.* London: Methuen, 1982.

Leakey, Richard, and Roger Lewin. *The Sixth Extinction: Patterns of Life and the Future of Mankind.* New York: Doubleday, 1995.

Mintzer, Irving M., ed. *Confronting Climate Change: Risks, Implications and Responses.* Cambridge: Cambridge University Press, 1992.

Moore, Thomas Gale. *Climate of Fear: Why We Shouldn't Worry about Global Warming.* Washington, D.C.: Cato Institute, 1998.

Stevens, William K. *The Change in the Weather.* New York: Delacorte Press, 1999.

Stommel, Henry, and Elizabeth Stommel. *Volcano Weather: The Story of the Year without a Summer, 1816.* Newport, R.I.: Seven Seas Press, 1983.

Upgren, Arthur. *Night Has a Thousand Eyes: A Naked-Eye Guide to the Sky, Its Science and Lore.* New York: Plenum, 1998.

Ward, Peter D. *The Call of Distant Mammoths: Why the Ice Age Mammals Disappeared.* New York: Copernicus, Springer Verlag, 1997.

INDEX

Thunderstorms, 73–74
Tidal waves, 175
Tornadoes, 74
Torricelli, Evangelista, xi
Transportation, cargo, and energy
    consumption, 192–193
Tree rings, 110–111, 122
Tropics
    fronts in, 58
    future expansion of, 185
    seasons and temperature variation,
        39–40, 167
    storms, 71–73
Troposphere, 17
Tunguska Event, 89
Twain, Mark, 18
Typhoons, 71

Underdeveloped world. See Third world
    countries
Uniformitarianism, 81–82
Uranus, 38

Vegetation
    and carbon dioxide, 143
    and carbon storage, 128–129
    and climatological history, 110–114
    deforestation, 142–143
    See also Rain forests
Venezuela, 39–40, 167, 169–170
Venus, 38, 40–41, 44–46, 53–54, 187
Vikings, 119–120
Volcanic eruptions and atmospheric
    cooling, 115–117, 138

Walton, Izaak, 63
Ward, Peter, 82
Water
    boiling point, difficulty of measuring, 25
    and carbon dioxide, 47–48
    cloud formation, 205–206
    compress, refusal to, 8

dew point. See Dew Point
evaporation and energy exchange,
    206–207
expansion when freezing, 146–147
and greenhouse gases, 134
as heat conductor, 32
humidity. See Humidity
precipitation. See Precipitation
salinity, 151
in the solar system, 21
vapor in air, 15–16. See also Humidity
See also Oceans
Weather
    defined, 7
    history of. See Holocene Epoch;
        Pleistocene Epoch
    mistaken beliefs about, 23–24
    movement of, 52–53, 56, 209–210
Weather lore
    animals, 66–67
    clouds, 59, 62
    fronts, 58–59
    holidays and the calendar, 60–61
    the moon, 50–51, 67–68
    rainbows, 56
    seasons, 60
    sky color, 55–56
    wind, 63–66
Wind
    chill, 20–21
    and the Coriolis Effect, 209–210
    and El Niño, 154–155
    energy generation, 175–176
    geostrophic, 55
    measurement of velocity, 27–28
    and ocean currents, 149–151
    weather lore, 63–66
Wisconsin Ice Age, 93–95
Wittmer, Rolf, 155
Wood, as energy source, 177

Zero, absolute, 10, 26